# VEHICLE GEARBOX NOISE AND VIBRATION

# Automotive Series

**Series Editor: Thomas Kurfess**

# VEHICLE GEARBOX NOISE AND VIBRATION

## MEASUREMENT, SIGNAL ANALYSIS, SIGNAL PROCESSING AND NOISE REDUCTION MEASURES

**Jiří Tůma**
**VŠB** *Technical University of Ostrava, Czech Republic*

*Registered office*
John Wiley & Sons Ltd, The Atrium, Southern Gate, Chichester, West Sussex, PO19 8SQ, United Kingdom

For details of our global editorial offices, for customer services and for information about how to apply for permission to reuse the copyright material in this book please see our website at www.wiley.com.

*Library of Congress Cataloging-in-Publication Data*

Tuma, Jiri (Engineer)
  Vehicle gearbox noise and vibration / by Jiri Tuma.
    1 online resource.
  Includes bibliographical references and index.
  Description based on print version record and CIP data provided by publisher; resource not viewed.
    ISBN 978-1-118-79761-7 (ePub) – ISBN 978-1-118-79762-4 (Adobe PDF) –
ISBN 978-1-118-35941-9 (cloth)    1. Motor vehicles–Transmission devices–Noise.    I. Title.
  TL262
  629.2′440288–dc23

                                                                        2013051076

A catalogue record for this book is available from the British Library.

ISBN: 978-1-118-35941-9

Set in 10/12pt Times by Aptara Inc., New Delhi, India
Printed and bound in Malaysia by Vivar Printing Sdn Bhd

1   2014

To Magda, Lucie, Peter and Eva
To the memory of my parents and brother

# Contents

# Series Preface

The gearbox is one of the most critical components of the automobile. Indeed, it is one of the most critical components of many mechanical systems including those used in aerospace, maritime, construction and agricultural systems, to name a few. The intricate combination of rotating gears, bearing and shafts interacting in a wide variety of modes results in a complex set of dynamics defining the performance of the gearbox. This performance directly drives the ability to transmit power from the engine to the wheels. However, these interactions can also result in a significant amount of vibration and noise which can affect ride comfort, systems performance and even the safety of the overall vehicle due to issues such as durability and fatigue.

The *Automotive Series* publishes practical and topical books for researchers and practitioners in industry, and postgraduate/advanced undergraduates in automotive engineering. The series covers a wide range of topics, including design, manufacture and operation, and the intention is to provide a source of relevant information that will be of interest and benefit to people working in the field of automotive engineering. *Vehicle Gearbox Noise and Vibration* is an excellent addition to the series focusing on noise and vibration issues stemming from the gearbox. The text provides an excellent technical foundation for noise and vibration analysis based on significant past research and development efforts, as do many texts in this area. What makes this text unique is that the author expands upon the classical analysis techniques to integrate and make use of the latest, state-of-the-art technologies and concepts that are in use today. Finally, throughout the book, real world examples are given to demonstrate the application of the various techniques in combination with each other, providing the reader with some excellent insight into what can be expected when employing the various noise and vibration concepts.

As is mentioned in the beginning of this preface, *Vehicle Gearbox Noise and Vibration* is part of the *Automotive Series*; however, gearboxes are found on a wide variety of other systems outside of the automotive sector. Thus, the concepts presented in this text are applicable across a wide variety of fields. Issues related to noise and vibration of gearbox components such as shafts, gears and bearings further extend the utility of the concepts presented to a wide variety of rotating systems such as turbo pumps, aircraft engines and power generation, to name a few. Furthermore, the pragmatic signal processing techniques that are presented in the text are applicable to any physical engineering system, making the utility of this book quite far reaching.

*Vehicle Gearbox Noise and Vibration* nicely integrates a set of topics that are critical to rotating systems. It presents some very pragmatic applications of those techniques with real-world examples demonstrating the implementation of the presented concepts. It is state-of-the-art, written by a recognized expert in the field and is a valuable resource for experts in the field. It is a welcome addition to the *Automotive Series*.

Thomas Kurfess
*January 2014*

# Preface

Many books deal with the calculation of gear geometry with respect to gear strength, selection of material, lubrication of teeth, alignment of gears and estimation of wear. However, there is less information available on how they are manufactured, and almost none is available on such details as the meshing cycle of teeth through the measurement of gearbox vibrations. The usual measurements of gearbox vibration and noise provide a frequency spectrum. The frequency spectrum does not give direct information about the meshing cycle. The first articles on the evaluation of toothmeshing in the time domain appeared about 25 years ago. In the frequency domain the responses of the loaded gears are separated from each other by different frequencies, but this does also have an impact on the time domain. The time course of vibrations becomes useful only when it is able to focus on a selected gear train and filters out the vibration responses of the other gear trains. This technique is known as synchronous filtration or synchronous averaging. Another name for it is signal enhancement. This method of signal processing requires the signal to be resampled synchronously with the rotational speed, which allows the angular vibrations during rotation to be determined.

A substantial part of this book discusses how the time domain analysis of the transmission unit is applicable to any rotating machine. This book also describes the practical measurement of angular vibration during rotation and how this is associated with the method of measuring transmission error of the gear train. However, many researchers, especially in the UK and the USA (for example D. Smith, R.G. Munro, D. Hauser) have already carried out this measurement.

The gearboxes of the vehicles do not operate at a constant speed or a constant load. The variable speed requires changes in the gear meshing frequencies to be tracked and spectral peaks which are excited by meshing gears or due to the resonance of the mechanical structure to be distinguised. This book describes how the run-up and coast down of machines can be analysed using the time-frequency representation a multispectrum. An alternative way to analyse the transient states is to use tracking filters such as quadrature mixing and the Vold-Kalman tracking filtration.

This book describes how to interpret the composition of the real cepstrum. The difference between the cepstrum of harmonics, odd harmonics and the set of harmonics which contain the sidebands of carrier components is demonstrated. The fundamental frequency of the harmonics and odd harmonics is related to the zero frequency, while the fundamental frequency of the harmonic components as sidebands is related to the carrier component frequency.

The main topic of the book is a description of the research work which was done to reduce the gearbox noise of a heavy-duty vehicle. There are two possible solutions for keeping a transmission unit quiet. Introducing an enclosure for preventing noise radiation is the easiest one, but it has consequences, for example, low efficiency and maintenance difficulties. The more sophisticated and much more efficient solution is based on solving the noise problem at the source. It means introducing improvement aimed at the gear design and manufacturing, which results in the greatest reduction of noise level as is shown.

The final chapter of the book describes the process of deciding how to proceed when using the most effective noise control measures. The ratio of radiated noise power of the individual units such as engine, gearbox, axles and tyres to the overall noise level during the pass-by noise test of the vehicle is analysed. Based on the resulting statistics the effect of limiting the deviations during production is estimated. The need to increase the stiffness of the transmission housing has been demonstrated by measuring the vibrations at the different gear ratios. The final decision was to change the contact ratio of the gears from low (LCR) to high (HCR). To keep the radiated noise under control, the effect of load, the gear contact ratio and the tooth surface modification on noise and vibration are illustrated by measurement examples giving an idea of how to reduce transmission noise.

In addition to describing the noise problems of the vehicle gearbox the book is also a textbook of signal processing. The chapter which deals with the demodulation of the modulated signals is universally applicable to the diagnostics of machines. In particular, the described methods for the measurement of angular vibrations are not that well known and are waiting for further applications. The book contains the first detailed description of the Vold-Kalman order tracking filter.

# Acknowledgements

I was first asked to write a book about gearbox vibration and noise by Debbie Cox at the 16th International Conference on Noise and Vibration, held in Krakow, Poland between 5–9 July 2009, where I presented a keynote lecture on this topic. To start with I was supervised by Debbie Cox and later by Tom Carter. I thank both for their help. I would like to mention all those who gave me the opportunity to work in the field of vehicle noise and vibration. Everything started when TATRA decided to solve the problem of noise from heavy-duty vehicles. My role in the project of the development of a quiet heavy-duty vehicle was to oversee noise and vibration measurements, signal processing, statistical investigation and the development of the software which supported measurements and evaluations. Fortunately, I could collaborate with very experienced designers, testing engineers and specialists working in technology and quality control of production.

I am grateful to Professor V.Moravec, the head of the team for developing the HCR gearing at TATRA at the beginning of the 1990s, for his help and valuable comments on gear design and accuracy. Not forgetting, fellow team members R. Kubena, V. Nykl and F. Sasin. Thanks to all.

Later, I started working at the VSB – Technical University of Ostrava, so I would like to thank colleagues from the Department of Mechanisms and Machine Parts and the Department of Control Systems and Instrumentation for cooperation in the development of methods for measuring transmission error. In particular, I must mention Professor Z. Dejl. For assistance with technical equipment there is Professor L. Smutný. Cooperation between the university and Tatra continues. Most valuable is the exchange of experiences with J. Jakubec, the chief designer of transmissions.

I am grateful to Professor M. J. Crocker, who, as editor-in-chief and co-author, invited me to publish a chapter in the Handbook of Noise and Vibration Control. This invitation helped me to start performing at international level.

Research work on the transmission error measurement that has been carried out at the VSB – Technical University of Ostrava, was supported by the Czech Science Foundation.

# 1

# Introduction

Various authorities aim to reduce the noise level in the environment by issuing requirements for the maximum noise level of critical noise resources. In transport, it is primarily motor vehicles which are subject to noise emission regulations. However, the strict limits cannot be introduced all at once, therefore the reduction is expected to be made gradually over at least 25 years. Newly manufactured vehicles which do not meet specified noise limits do not obtain permission to operate on public roads. Motor vehicle manufacturers have been given sufficient time to implement noise reduction innovations. The time line for noise limits for cars and trucks with an engine power of 150 kW and more is shown in Figure 1.1. Data was taken from the final report of the working party on noise emissions of road vehicles. The arrow pointing at 1985 indicates that in the EU there was a change in measuring procedure. For trucks, this corresponded to 2–4 dB of stricter requirements on top of the other changes; but for cars it corresponded to approximately 2 dB of less stringent requirements.

There is an international standard for the measurement of noise emitted into the environment. Details will be discussed in the last chapter of the book. For now, it is sufficient to note that under certain conditions the Sound Level Metre measures the maximum of the sound pressure level at the point which is at a distance of 7.5 m from the centreline of the track of the vehicle and 1.5 m above the road surface. The same sound pressure level is measured in the USA at the distance which is twice as far away, so limits for this country were raised to about 6 dB in the graph in Figure 1.1. This measurement relates to pass-by noise. The noise level in the vehicle cabin is a separate factor.

So began a race against time for manufacturers of heavy trucks. The sound pressure limit of 84 dB was not difficult to meet. But to produce a heavy-duty vehicle of 80 dB required changing the design. Transmissions can be put into an enclosure with a small reduction of 4 dB in the level of radiated noise or it is possible through a fundamental change in the parameters of gears [2, 3]. This book describes the difficult development which led to a substantial reduction in noise transmission by improving the design of gears. The theme of the book does not address the design, but describes the methods of measurement and signal processing which helped to determine the effect of design modifications or just to verify the correctness of the decision.

*Vehicle Gearbox Noise and Vibration: Measurement, Signal Analysis, Signal Processing and Noise Reduction Measures,*
First Edition. Jiří Tůma.

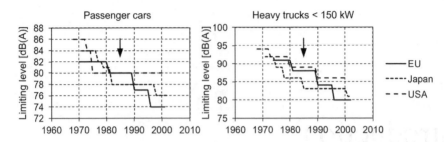

**Figure 1.1**   Development of vehicle noise emission limits over the years [1].

## 1.1   Description of the TATRA Truck Powertrain System

The theory of signal processing is illustrated by examples of the measurement of noise and vibration of the gearbox of the TATRA trucks. It is therefore appropriate to describe the transmission of these vehicles in detail. The truck powertrain system consists of the engine, gearbox, differentials and axles. All these units contain gears. Due to the high rotational speed and transferred torque, gears in a gearbox and axles play a key role in emitting noise. All gears in the TATRA gearbox are of the helical type and the gears in the axles are of the spiral bevel type. The problem of axle noise is serious, but this book does not propose to cover this area of research in detail. In Chapter 7 a method that enables the contribution of the noise level emitted by the axle to the overall noise level of the vehicle to be evaluated is discussed.

There are a number of gears which rotate in the truck as is shown in Figure 1.2. These include the timing gears of a diesel engine, but these are not a source of serious noise. The

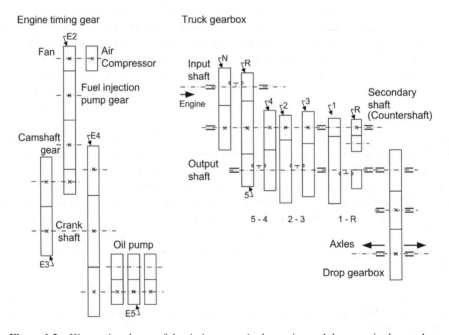

**Figure 1.2**   Kinematic scheme of the timing gears in the engine and the gears in the gearbox.

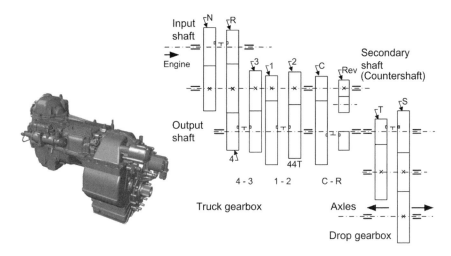

**Figure 1.3**   Kinematic scheme of the newest model of the TATRA gearbox.

main source of noise which is produced by gears is the transmission unit. The older gearbox unit, including a drop or secondary gearbox, is in the left of Figure 1.2.

The secondary gearbox is sometimes called the drop gearbox due to the fact that this gearbox reduces the rotational speed. In the case of the TATRA trucks, the drop gearbox transfers power to the level of the central tube, which is the backbone of the chassis structure. The main gearbox comprises two stages and has five basic gears and reverse. As all the basic gears are split (R, N) the total number of the basic gears is extended to ten forward and two reverse gears. The gears are designated by a combination of the number character (1 up to 5 or 6) and letter (R or N), for example '3N'. According to the EEC regulations valid at the beginning of the 1990s, the basic gears selected for the pass-by tests are 3, 4 and 5. The drop gearbox is either the compound gear train with an idler gear or the two-stage gearbox, extending the number of gears to 12. TATRA does not use a planetary gearbox as the drop gearbox.

A kinematic scheme of the newest model of the TATRA gearbox is shown in Figure 1.3. The drop gearbox has two gear ratios in contrast to the old model of the gearbox. As is evident from the kinematic schemes both transmissions are manual and all the gears are synchronised.

## 1.2   Test Stands

The operating conditions of gearboxes can be simulated using test rigs to drive the gearbox in a similar way to the pass-by noise test. The configuration of the closed loop is energy saving. With the use of an auxiliary planetary gearbox the torque is inserted in the closed circuit while an auxiliary electric motor spins the system at the operational speed. Power, which is the product of angular velocity and torque, then circulates inside the loop. If the auxiliary transmission adapts to different variants of the gearbox under test, then the power consumption of the test rig increases for example, up to 40% of the power that circulates in a closed loop.

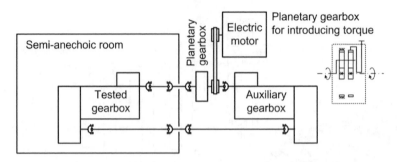

**Figure 1.4**   Closed loop test rig for testing noise in semi-anechoic room.

An example of a closed circuit arrangement is shown in Figure 1.4. According to current standards for testing the radiated sound pressure level the volume of the chamber should be at least 200 times larger than the volume of the test gearbox. Microphones are placed on the sides of the gearbox in the direction of the truck movement at a distance of 1 m. Accelerometers that are attached on the surface of the gearbox housing near the shaft bearings can provide extensive information about the noise sources. A tacho probe, generating a string of pulses, is usually employed to measure the gearbox-primary-shaft rotational speed. A sensor for measuring the torque is also inserted into the closed loop.

In contrast to the open loop test stand, the back-to-back test rig configuration saves drive energy. The torque to be transmitted by the gearbox is induced by a planetary gearbox. The gearbox under testing is enclosed in a semi-anechoic room with walls and ceiling absorbing sound waves and a reflective floor. The quality of the semi-anechoic room is of great importance for the reliability of the results. The reverberation time should be less than is required in the frequency range from at least 200 to 3 kHz. The input shaft speed is slowly increased from a minimal to maximal RPM while the gearbox is under a load corresponding to full vehicle 'acceleration'. To simulate the gearbox operational condition during deceleration the noise test continues to slowly decrease from a maximal to minimal RPM.

The configuration for measuring an open loop is shown in Figure 1.5. Noise is measured in the open field with two microphones that are located in an anechoic chamber. Because the eddy current brake is used, it is necessary to use an auxiliary gearbox to increase the speed at which this type of the brake is able to effectively load the gearbox by a torque.

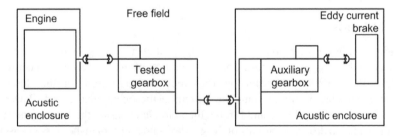

**Figure 1.5**   Open loop test rig for testing noise in free field.

# References

[1] Sandberg, U. (2001) Noise emissions of road vehicles effect of regulations, Final Report 01-1. I-INCE working party on noise emissions of road vehicles (WP-NERV), International Institute of Noise Control Engineering.

[2] Arenas, J.P. and Crocker, M.J. (2010) Recent trends in porous sound-absorbing materials. *Sound and Vibration*, **44**(7), 12–17.

[3] Zhou, R. and Crocker, M.J. (2010) Sound transmission loss of foam-filled honeycomb sandwich panels using statistical energy analysis and theoretical and measured dynamic properties. *Journal of Sound and Vibration*, **329**(6), 673–686.

References

# 2

# Tools for Gearbox Noise and Vibration Frequency Analysis

The signal $x(t)$ is a real or complex function of continuous time $t$. The other definition points to the fact that the signal contains information which transmits from the source to the receiver. But one of the signal types called a white noise does not formally contain any information. White noise is a totally random signal and the present samples do not depend on the past samples in any way. Signals describe the noise and vibration as time processes, and have common characteristics. This chapter deals with the theory of digitisation of analogue signals and different methods of signal processing in the time and frequency domain.

## 2.1 Theory of Digitisation of Analogue Signals

### 2.1.1 Types of Signals

Now we turn attention to the types of signals. The basic types of signals are deterministic and stochastic. There are deterministic periodic or non-periodic signals. The non-periodic signals can be broken down into almost periodic or transient signals. Simple tone seems to be deterministic, while multi-tonal sound seems to be stochastic (random). The signals in practice are a mixture of deterministic and random components. Further subdivision is shown in Table 2.1.

Deterministic signals are defined as a function of time while random signals can be defined in terms of statistical properties. The deterministic signals can be predicted, while random signals, which have instantaneous values, are not predictable. The theory of random signals is based on the following system of naming. Names of random variables and processes are Greek letters:

$$\text{variables: } \xi, \varepsilon, \ldots \text{ processes: } \xi(t), \varepsilon(t), \ldots$$

while the measured waveforms, called realisations, are identified as Latin letters:

$$x(t), y(t), \ldots$$

*Vehicle Gearbox Noise and Vibration: Measurement, Signal Analysis, Signal Processing and Noise Reduction Measures,*
First Edition. Jiří Tůma.
© 2014 John Wiley & Sons, Ltd. Published 2014 by John Wiley & Sons, Ltd.

**Table 2.1**   Types of signals.

| Deterministic | | | | Random (stochastic) | | |
|---|---|---|---|---|---|---|
| Periodic | | Non-periodic | | Stationary | | Non-stationary |
| Sinusoidal | Complex periodic | Almost periodic | Transient | Ergodic | Non-ergodic | Special classification |

A probability density function reflects the basic properties of random variables. Probability that a random variable $\xi$ belongs to the interval of values greater than $x$ and less than $x + \Delta x$ is proportional to the interval of the length $\Delta x$

$$P\{x < \xi \leq x + \Delta x\} = p(x)\,\Delta x \tag{2.1}$$

The coefficient of proportionality $p(x)$ is denoted as a probability density function (pdf). The function $p(x)$ is one-dimensional. There are also two and more dimensional pdf $p(x_1, x_2, \ldots)$, called joint probability. Between random variables and random signals (processes) is the relationship as it is documented in Figure 2.1. Values of the random process at time $t_1$ become a random variable.

The probability density function is a basic property of random signals for the definition of the mean value $\mu$ and variance $\sigma^2$

$$\mu = \mathrm{E}\{\xi\} = \int_{-\infty}^{+\infty} x p(x)\,dx$$

$$\sigma^2 = \mathrm{var}\{\xi\} = \mathrm{E}\{(\xi - \mu)^2\} = \int_{-\infty}^{+\infty} (x - \mu)^2\, p(x)\,dx. \tag{2.2}$$

Square root of the variance is a standard deviation $\sigma = \sqrt{\mathrm{E}\{(\xi - \mu)^2\}}$.

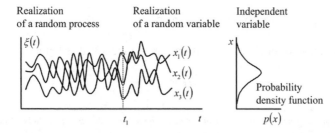

**Figure 2.1**   Relationship between random processes and variables.

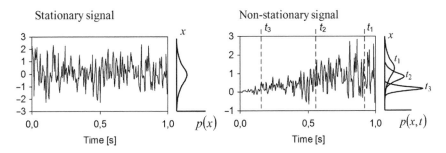

**Figure 2.2** Stationary and nonstationary signals.

An important property of random processes is stationarity, which is defined using the dependence of the probability density function on time. A stationary signal is a stochastic signal or process whose joint probability distribution does not change when shifted in time or space. As a result, parameters such as the mean and variance, if they exist, also do not change over time. The visual difference between the stationary and nonstationary signal is obvious from Figure 2.2.

The basic property of a stationary continuous signal $x(t)$ is that one-dimensional pdf and consequently the mean value of the random signal is independent of time

$$p\left(x_1, t_1\right) = p\left(x_1\right). \tag{2.3}$$

and the two-dimensional pdf depends only on the time difference $t_1 - t_2$ which is the time that elapses between these two time instants

$$p\left(x_1, x_2, t_1, t_2\right) = p\left(x_1, x_2, t_1 - t_2\right) \tag{2.4}$$

To understand the calculation of mean values and variances it is necessary to introduce the concept of ergodic processes or signals. An ergodic process is one which is complying with the ergodic theorem. This theorem allows the time average of a signal to be equal to the ensemble average. In practice this means that statistical sampling can be performed at one instant across a group of identical signals or sampled over time on a single signal with no change in the measured result.

$$\bar{x} = \int_{-\infty}^{+\infty} x p\left(x\right) \mathrm{d}x = \lim_{T \to +\infty} \frac{1}{T} \int_{-T/2}^{+T/2} x\left(t\right) \mathrm{d}t. \tag{2.5}$$

This assumption is crucial to the process of measurements, because it allows practical results to be obtained.

## 2.1.2  Normal Distribution

The normal distribution of a random variable is defined by the formula

$$p(x) = \frac{1}{\sigma\sqrt{2\pi}} \exp\left(-\frac{1}{2}\left(\frac{x-\mu}{\sigma}\right)^2\right) \qquad (2.6)$$

where $\mu$ is the mean value and $\sigma$ is the standard deviation of the mentioned random variable which were defined above. For a random variable $\xi$ the notation is as follows:

$$p(\xi) \sim N\left(\mu, \sigma^2\right).$$

The multivariate normal distribution of the random vector is defined by the formula

$$p(\mathbf{x}) = \frac{1}{(2\pi)^{\frac{k}{2}} |\mathbf{P}|^{\frac{1}{2}}} \exp\left(-\frac{1}{2}(\mathbf{x}-\boldsymbol{\mu})^T \mathbf{P}^{-1}(\mathbf{x}-\boldsymbol{\mu})\right) \qquad (2.7)$$

where $\mathbf{x}$ is a vector of variables of the size $k \times 1$, $\boldsymbol{\mu}$ is a vector of mean values and $\mathbf{P}$ is a covariance matrix which is a positive definite symmetric matrix

$$\begin{array}{l} \mathbf{x} = \left[x_1, x_2, x_3, \ldots, x_k\right]^T \\ \boldsymbol{\mu} = \left[\mu_1, \mu_2, \mu_3, \ldots, \mu_k\right]^T \end{array} \quad \mathbf{P} = \begin{bmatrix} \sigma_1^2 & \cdots & \sigma_{1,k} \\ \vdots & \ddots & \vdots \\ \sigma_{k,1} & \cdots & \sigma_k^2 \end{bmatrix}. \qquad (2.8)$$

For a random variable $\xi$ the notation is as follows: $p(\mathbf{x}) \sim N(\boldsymbol{\mu}, \mathbf{P})$.

## 2.1.3  Mean Value and Standard Deviation (RMS) of a General Signal

In the case of signals, it is easier to calculate the time mean value from a sufficiently long record of a signal than from statistical data consisting of many realisations. Variance of an ergodic signal is given by the formula which is the similar to Eq. (2.5)

$$\sigma^2 = \int_{-\infty}^{+\infty} (x - \bar{x})^2 p(x)\,dx = \lim_{T \to +\infty} \frac{1}{T} \int_{-T/2}^{+T/2} (x(t) - \bar{x})^2\,dt. \qquad (2.9)$$

Signal processing for signals such as sound pressure, vibration, voltage and electrical current is based on the calculation of the root mean square (RMS or rms). This abbreviation RMS describes the order of writing the detailed parts of the formula

$$RMS = \sqrt{\frac{1}{T} \int_0^T x(t)^2\,dt}. \qquad (2.10)$$

Because the mean value of signals such as sound pressure and vibration are equal to zero then the standard deviation and RMS are numerically identical.

If we process electrical signals, such as voltage $u(t)$ and current $i(t)$, then we can define instantaneous power $p(t) = u(t)\,i(t)$. If the electric current and voltage is related to the resistance $R$ of the conductor, then the instantaneous power is given by the formula.

$$p(t) = Ri(t)^2 = \frac{u(t)^2}{R}. \tag{2.11}$$

In both cases the instantaneous power is proportional to the square of the current or voltage. The resistance R can be regarded as a scaling factor. Therefore it is useful to introduce the general power of the signal as the square of the signal. This power can be either instantaneous or an average of the instantaneous power for the selected time interval. The mean power PWR of the signal in a certain time interval of the length $T$ is given by

$$PWR = \frac{1}{T} \int_0^T x(t)^2 \, dt = RMS^2. \tag{2.12}$$

As regards the units, the sound pressure, acceleration and the velocity have the units Pa, m/s$^2$ and m/s, respectively. The signal power of the sound pressure, acceleration and the velocity have the unit as follows Pa$^2$, (m/s$^2$)$^2$, (m/s)$^2$. A signal energy is the sum or integer with respect to time of the instantaneous power $\int x(t)^2 \, dt$. The unit of the signal energy is multiplied by seconds, for example, Pa$^2$s, (m/s$^2$)$^2$s.

### 2.1.4 Covariance

The covariance between two real-valued random variables $x$ and $y$ is defined by the formula

$$\mathrm{cov}(x, y) = E\left\{(x - E\{x\})(y - E\{y\})\right\}. \tag{2.13}$$

If both the random variables $x$ and $y$ are identical, $x = y$, then the covariance becomes the variance, $\mathrm{cov}(x, x) = \mathrm{var}(x)$. The covariance between random vectors $\mathbf{X}$ and $\mathbf{Y}$ of dimension $(m \times 1)$ and $(n \times 1)$, respectively, is a matrix defined by

$$\mathrm{cov}(\mathbf{X}, \mathbf{Y}) = E\left\{(\mathbf{X} - E\{\mathbf{X}\})(\mathbf{Y} - E\{\mathbf{Y}\})^T\right\} = E\left\{\mathbf{X}\mathbf{Y}^T\right\} - E\{\mathbf{X}\}\,E\{\mathbf{Y}\}^T. \tag{2.14}$$

There is a conflict in the notation of the variance and covariance random vector. Some statisticians use this notation

$$\mathrm{var}(\mathbf{X}) = \mathrm{cov}(\mathbf{X}) = E\left\{(\mathbf{X} - E\{\mathbf{X}\})(\mathbf{X} - E\{\mathbf{X}\})^T\right\} = E\left\{\mathbf{X}\mathbf{X}^T\right\} - E\{\mathbf{X}\}\,E\{\mathbf{X}\}^T. \tag{2.15}$$

## 2.1.5   Mean Value and Standard Deviation (RMS) of a Sinusoidal Signal

A sinusoidal signal with the amplitude $A$ and with the period of the length $T$ is defined by the following formula

$$x(t) = A \cos\left(\frac{2\pi}{T}t\right). \tag{2.16}$$

The mean value of the sinusoidal signal over a single full period is zero

$$\bar{x} = \frac{1}{T}\int_0^T x(t)\,dt = 0. \tag{2.17}$$

The variance and standard deviation of the sinusoidal signal over a single full period is given by

$$\sigma^2 = \frac{1}{T}\int_0^T (x(t))^2\,dt = \frac{1}{T}\int_0^T \left(A\cos\left(\frac{2\pi}{T}t\right)\right)^2\,dt = \frac{1}{T}\int_0^T \frac{A^2}{2}\left(1+\cos\left(\frac{4\pi}{T}t\right)\right)\,dt = \frac{A^2}{2}$$

$$\sigma = \text{RMS} = \frac{A}{\sqrt{2}}. \tag{2.18}$$

The value of RMS of the harmonic signal is approximately equal to 70% of its amplitude and the amplitude of this signal is the 1.4-multiple of RMS.

## 2.1.6   Digitalisation of Signals

All measurement data are processed on digital computers. Analogue devices belong to history, with a few exceptions where analogue filters are used. The process of converting of an analogue signal into a digital signal is called digitalisation. The digital signal is a function of discrete time or a sequence of sample. There are two issues in digitalisation:

- sampling
- quantising.

First, we discuss sampling. Sampling of an analogue continuous time signal $x(t)$ produces a sequence of samples at discrete time $t_n = nT_S$, where $T_S$ is a sampling interval and $n$ is an integer number. Sampling at equi-spaced time instants is still the most widely used technique. The sequence of samples may be denoted either as an indexed variable $x_n$ or as a function $x(nT_S)$ or just $x(n)$ of this index $n$. Within the scope of this book, vibration, noise and any other variable which depends on time are considered as a signal.

Sampled signals are predetermined for processing on digital computers. The following formulas for calculation of basic parameters, such as the mean value $\bar{x}$ and variance $s^2$ of a digitalised signal $x_n, n = 0, \ldots, N-1$, where $N$ is a sample number, assume that the result of a sampling process is sequences corresponding to ergodic signals. The mean value and variance are computed by the following formulas:

$$\bar{x} = \frac{1}{N} \sum_{n=0}^{N-1} x_n$$

$$s^2 = \frac{1}{N} \sum_{n=0}^{N-1} (x_n - \bar{x})^2. \tag{2.19}$$

The square root of the variance is a standard deviation. Assuming that the mean value $\bar{x}$ of a signal is zero, the result of the calculation of the standard deviation is RMS

$$RMS = \sqrt{\frac{1}{N} \sum_{n=0}^{N-1} x_n^2}. \tag{2.20}$$

Noise and vibration signals have zero mean value therefore the standard deviation and RMS is numerically identical.

Quantising is a part of the analogue to digital (AD) conversion which results in the rounded output value of the actual input value of the signal at discrete time. The consequence of rounding is a presence of additional quantising noise in signals, see Figure 2.3.

Assume that the AD converter (ADC) produces a digital output of the integer type. If the digital value is represented by an $m$-bit integer then the unsigned output ranges from 0 to $2^m - 1$ and the signed output ranges from $-2^{m-1}$ to $+2^{m-1} - 1$. An integer number at the output of ADC, which is denoted by $k$ for instance, theoretically corresponds to the input value of the signal ranging between $k - 0.5$ and $k + 0.5$. If rounding is unbiased, then the mean value of the rounding error is zero. The rounding error has a uniform probability density

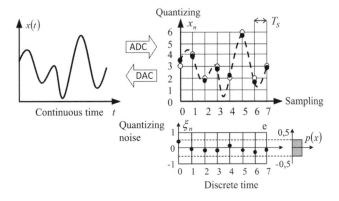

**Figure 2.3**  Sampling and quantising an analogue signal.

function $p(x)$ on the interval ranging from $-0.5$ to $+0.5$. The variance of the quantising noise is calculated according to formula (2.2) as follows

$$\sigma_\xi^2 = \int\limits_{-\infty}^{+\infty} x^2 p(x)\, dx = \int\limits_{-0.5}^{+0.5} x^2\, dx = \frac{1}{12} \qquad (2.21)$$

The standard deviation of the quantising noise is equal to $\sigma_\xi = 1/\sqrt{12}$.

## 2.1.7   Signal-to-Noise Ratio

The AD conversion corrupts the digitised signal by rounding noise which is added to the original analogue signal. The rounding noise is also called background noise and the original signal is considered as meaningful or useful information which is transferred via the AD converter. The ratio of the useful signal power $S_{PWR}$ to noise power $N_{PWR}$ is a parameter to assess the AD converter. Decibels are used due to a large range of the aforementioned ratio. Signal-to-Noise Ratio in dB is defined by the following formula

$$S/N_{dB} = 10 \log\left(\frac{S_{PWR}}{N_{PWR}}\right) \qquad (2.22)$$

The power PWR is equal to the square of RMS. An alternative definition of the Signal-to-Noise Ratio is given by the formula

$$S/N_{dB} = 20 \log\left(\frac{S_{RMS}}{N_{RMS}}\right) \qquad (2.23)$$

For the $m$-bit AD converter, the maximum amplitude of the harmonic signal without crossing clipping level is equal to $2^m/2 = 2^{m-1}$ and the corresponding RMS of the signal $S_{RMS}$ is equal to $2^{m-1}/\sqrt{2}$. The value of $N_{RMS}$ for the quantising noise is given by Eq. (2.21). The standard deviation of quantising noise $\sigma_\xi = 1/\sqrt{12}$ is only another designation of the same variable. After substituting we obtain the formula for calculating the signal-to-noise ratio of just putting the number $m$ of bits in the formula as follows

$$S/N_{dB} = 20 \log\left(\frac{2^{m-1}/\sqrt{2}}{\sigma_\xi}\right) = 20\left[(m-1)\log(2) + \frac{1}{2}\log(6)\right] \qquad (2.24)$$

$$S/N_{dB} = 6.02m + 1.76\,\text{dB}$$

Signal-to-noise ratio is approximately equal to six times the number of bits of the A/D converter. The effective number of bits of the converter is usually less than the number of bits of the digital data at the converter output. Table 2.2 shows how the signal to noise ratio

**Table 2.2** Signal-to-Noise Ratio as a function of the ADC bits.

| Number of ADC bits | 4 | 8 | 10 | 12 | 14 | 16 | 18 | 20 | 22 | 24 |
|---|---|---|---|---|---|---|---|---|---|---|
| Maximum of $S/N_{dB}$ | 26 | 50 | 60 | 72 | 84 | 96 | 108 | 120 | 132 | 144 |

depends on the number of bits of the AD converter. Previously, there were signal analysers on the market with the AD converters that had 14 or 16 bits. Today, the standard is 24 bits.

The effective number of bits (ENOB) is usually smaller than the catalogue value. The reasons for this include: harmonic distortion, cross-talk, power supply imperfection, clock circuits, EMC coupling between circuits, non-linearity and aliasing originating from signal components of frequencies higher than half the sampling frequency.

## 2.1.8 Sampling as a Mapping

This part of the fundamental description of signal processing is not required if you skip the theory and formulas for continuous time and you are only interested in the sampled signals. Anyone who wants to understand the context may be familiar with the details of the general theory. Sampling may be considered as a mapping

$$\text{Sampling: } x(t) \rightarrow \left[x_0, x_1, x_2, \ldots\right]^T. \tag{2.25}$$

In fact this mapping is the function that determines how to extract a sequence of samples from the continuous-time signal as shown in Figure 2.4.

Now we turn our attention to the creation of continuous-time functions from the sequence of samples. For this purpose we need a special function, called the Dirac delta function. The Dirac delta is not strictly a function. It is called a distribution. The delta function can be viewed as a limit of the sequence of functions $\delta(t) = \lim_{n \to \infty} \delta_n(t)$, where $\delta_n(t)$ is sometimes called a nascent delta function. This limit is in the sense that

$$\lim_{n \to \infty} \int_{-\infty}^{+\infty} \delta_n(t-a)f(t)\,dt = f(a) \tag{2.26}$$

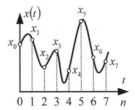

**Figure 2.4** Sampling of an analogue signal.

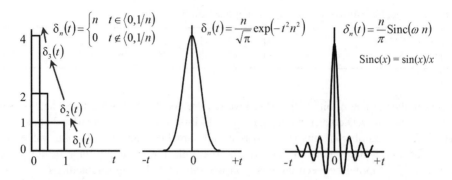

**Figure 2.5**   Sequence of the nascent delta functions tending to Dirac function.

The filtration effect of the Dirac delta function is part of the integrated function

$$\int_{-\infty}^{+\infty} \delta(t-a)f(t)\,dt = f(a) \tag{2.27}$$

Note that examples of the nascent functions in Figure 2.5 are different, but the effect on integration (2.27) is the same

The function of a continuous-time can be created from a sequence of samples in the following way

$$x^*(t) = \sum_{n} x_n \delta(t - nT_S) \tag{2.28}$$

where $T_S$ is a sampling interval for the equi-spaced sampled data. The sampling frequency (rate) $f_S$ measured in hertz (Hz) or in samples per second is the reciprocal value of the sampling interval. As will be explained below, the function $x^*(t)$ can reconstruct to the original function $x(t)$, provided that the Nyquist-Shannon sampling theorem is fulfilled. Therefore we can substitute $x(t)$ in the transformation formulas for the Fourier transform by the following term

$$\sum_{n} x(t)\,\delta(t - nT_S) = \sum_{n} x_n \delta(t - nT_S) \rightarrow x(t) \tag{2.29}$$

## 2.2   Nyquist-Shannon Sampling Theorem

This theorem is named after famous scientists contributing to the development of the theory of information, telecommunications and signal processing. However, for our purposes the name will be simplified to the sampling theorem. Selecting the sampling interval is not arbitrary, but is influenced by the requirement of signal reconstruction from a sequence of equally spaced samples in time. As will be shown below, the sampling interval of the sinusoidal signal has to be adapted to the length of period of this sinusoidal signal.

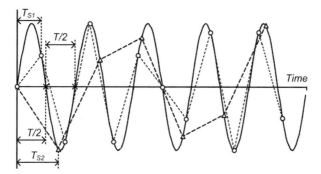

**Figure 2.6**   Sampling an analogue signal.

First, we will deal with the unique reconstruction of the mentioned sinusoidal signal, which is replaced by a sequence of equally spaced samples in time with a sampling interval $T_S$. An example is shown in Figure 2.6. The length of one waveform of the sinusoidal signal is designated by $T$. The frequency of the original sinusoidal signal can be determined from the number of crossings at the zero level. For counting the zero crossings the points of the sampled signal are connected by a line segment. As shown in Figure 2.6 we chose for sampling two intervals that are designated by $T_{S1}$ and $T_{S2}$. Sampling interval $T_{S1}$ is less than half the length of one waveform of the sinusoidal signal and the second sampling interval $T_{S2}$ is greater than the length of the sinusoidal signal period. Then the frequency of the signal period $T$ is designated by $f = 1/T$ and the sampling frequencies are designated by $f_{S1} = 1/T_{S1}$ and $f_{S2} = 1/T_{S2}$.

It is clear that only in the case of $T_{S1} < T/2$ is the number of zero crossings of the sampled signal equal to the number of zero crossings of the original sinusoidal signal, while sampling according to the case of $T_{S2} > T/2$ will change the frequency of the sampled signal. This phenomenon is called aliasing. The sampling frequency $f_S$ which is maintaining the sinusoidal signal frequency $f$ has to govern the following rule called the sampling theorem

$$T_S < T/2 \Rightarrow f_S > 2f. \tag{2.30}$$

Half the sampling frequency $f_S/2$ is called the Nyquist frequency.

It is clear that for a given frequency the values of two adjacent samples can be determined by the amplitude and phase of a sinusoidal signal. The reconstruction of the mentioned signal is possible only if sampling is subjected to compliance with the sampling theorem.

In other words, sampling with a possible reconstruction of the original signal is conditioned by the fact that the sampled signal does not contain a sinusoidal component with a frequency greater than half the sampling frequency. Reconstruction can also consist of determination of the frequency, phase and amplitude of any sinusoidal components of the analysed signal. In this sense the reconstruction means the ability to perform frequency analysis without mistakes caused by aliasing.

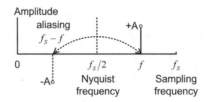

**Figure 2.7**   Aliasing.

## 2.2.1   Antialiasing Filter

Aliasing occurs when the frequency $f$ belongs to the interval $f_S/2 < f < f_S$. Assume that the signal $\sin(2\pi ft)$ is sampled at time instants indexed by $k = 0, 1, 2, \ldots$ and the frequency of the sampled signal is less than the sampling frequency. If the sampling frequency $f_S$ does not satisfy the sampling theorem and is greater than the Nyquist frequency then the frequency difference $f_S - f$ is less than the Nyquist frequency.

$$\sin\left(2\pi\frac{f}{f_S}k\right) = \sin\left(2\pi\frac{f - f_S + f_S}{f_S}k\right) = \sin\left(2\pi k - 2\pi\frac{f_S - f}{f_S}k\right)$$

$$= \sin(2\pi k)\cos\left(2\pi\frac{f_S - f}{f_S}k\right) - \cos(2\pi k)\sin\left(2\pi\frac{f_S - f}{f_S}k\right)$$

$$= -\sin\left(2\pi\frac{f_S - f}{f_S}k\right). \tag{2.31}$$

Graphical representation of a sinusoidal signal of the frequency $f$ which is sampled incorrectly with the sampling frequency $f_S$, seems to be after connecting points in the line chart as a signal whose frequency $f_S - f$ is lower than it was originally and has the opposite phase. The amplitude and frequency of an incorrectly sampled signal and its aliasing signal are shown in Figure 2.7. In the particular case of equality of the sampling frequency and frequency of a sinusoidal signal $f_S = f$ the result of the sampling process is a repeating constant value.

Requirements of the sampling theorem imply a need to adapt the sampling frequency to the largest frequency of the input signal or to adjust the frequency range of the signal to the possible sampling frequency of the given AD converter. The second solution, which is based on adjusting the frequency range of the analogue signal, implies the use of the low pass filter for this signal prior to connecting it to the input of the AD converter as it is shown in Figure 2.8. The analogue filter prevents the aliasing effect; therefore it is called an antialiasing filter.

The antialiasing analogue filter is of the type of the low-pass. According to Figure 2.8 it is clear that between the pass band and the stop band frequency range there is a transition band in the frequency response of the filter. Since it is impossible to design an analogue low pass filter without the transition band the frequency range of frequency analysis is limited to the frequency $f_{MAX}$, which coincides with the cut-off frequency $f_{CUT-OFF}$ of the LP filter.

The theoretical frequency range of signal analysis is equal to half the sampling frequency. Due to the presence of the transition band of the antialiasing filter the frequency range of

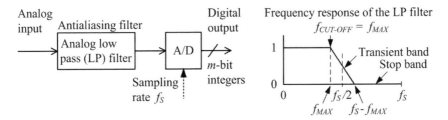

**Figure 2.8**   Antialiasing filter.

analysis is limited to $f_{MAX}$. For signal analysers which are produced by the Bruel and Kjaer Company the maximum frequency is as follows

$$f_{MAX} = f_S/2.56 \qquad (2.32)$$

A waveform of a sinusoidal signal requires on average 2.56 samples per period which are sampled by a constant frequency.

The analogue antialiasing filter is used for digitalising the signal at the highest sampling rate. Lower sampling rates are obtained by decimation of the primary digitalised signal. The process of decimation includes a digital low pass filtration.

The signal-to-noise ratio of a 14-bit AD converter is 84 dB, which means that the RMS of the quantising noise is about 84 dB lower than the maximum amplitude of the meaningful sinusoidal signal at the input of the antialiasing filter. Greater attenuation for a 14-bit convertor is not required due to rounding noise. This attenuation must be achieved in the transition band of the filter ranging from the frequency $f_{MAX}$ to $1.56 f_{MAX}$. The roll-off of the low pass filter has to be about 120 dB per octave.

Increasing the signal to noise ratio can be ensured by decreasing the frequency range of measurements with the use of a bandpass filter. The effect of narrowing the original frequency range of $f_{MAX}$ with the use of the mentioned filter of the bandwidth $BW$ is the following $10 \log \left( f_S/2BW \right)$. An overview of the properties of the basic types of AD converters on the market today is provided in Table 2.3. The table provides information about a typical accuracy resulting from the number of bits and also about whether you need an external antialiasing filter. The delay of conversion is important only for use in control loops.

Successful measurement of sound and vibration an A / D converter can be used with a sufficiently high signal-to-noise ratio. Furthermore, a converter of the appropriate type should

**Table 2.3**   Types of the AD converters.

| ADC principle | Number of bits | Antialiasing filter | Delay |
|---|---|---|---|
| Successive approximation | Max 12 (14) | Required | Low (depends on the bit number) |
| Sigma-delta | Up to 24 | Not required (a part of the converter) | Large (depends on LP filter order) |
| Flash | 8 (10) | Required | Negligible |

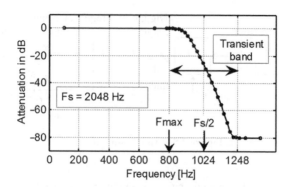

**Figure 2.9** Frequency properties of the antialiasing filter of BK PULSE 3109 [5].

include an antialiasing filter to measure signals without aliasing. The professional filter of this type includes signal analysers supplied by the Bruel and Kjaer Company, for example PULSE 3109. The frequency response of the antialiasing filter of this signal analyser is shown in Figure 2.9. The frequency response of the antialiasing filter was measured at a much lower frequency range than the highest frequency range of the analyser which is in fact equal to 25.6 kHz at the sampling frequency of 65 536 Hz. The lower frequency ranges are repeatedly derived by skipping every other sample. Before decimation of the sequence the sampled signal has to be low pass filtered with the use of a digital filter so as not to disturb the rule of the correct sampling. The frequency response shows the effect of both the analogue and digital antialiasing filters.

## 2.2.2   Sound and Vibration Measuring Chain

This book deals with the measurement and assessment of noise and vibration of transmission systems of vehicles, particularly gearboxes and the gear axis systems. An arrangement of the typical measuring chain for sound and vibration is shown in Figure 2.10.

Sound pressure and vibration in either acceleration or velocity or displacement are measured using transducers. Preamplifiers called signal conditioning are a part of the transducers or are situated close to them. The coaxial cables connect the preamplifiers to the ADC input. Voltage inputs called direct also belong with ADC inputs. These components of the measuring chain are in Tables 2.4 and 2.5.

An acronym BNC is formed from the initial letters of the pattern which is Bayonet and the names of the inventors Neill-Concelman. The TNC connector is a threaded version of the BNC connector. LEMO is the name of an electronic and fibre optic connector manufacturer, based in Switzerland.

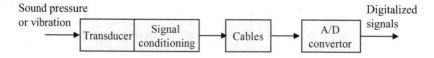

**Figure 2.10**   Measuring chain for sound and vibration.

**Table 2.4**   Transducers for sound and vibration measurements.

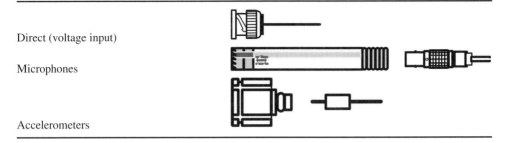

Direct (voltage input)

Microphones

Accelerometers

A/D converters are a part of the signal analysers or multifunction cards. Multichannel signal analysers have parallel inputs – each measuring channel has its own A/D converter. All the converters are triggered simultaneously in the same time. The multifunction cards work with a multiplexor which connects signals to the A/D converter. There is only one A/D converter on the multifunction card. If the sample-and-hold circuit is not used, then there is a time lag between the successively measured samples. The multifunction cards of low price are not equipped with an antialiasing filter except the cards for dynamic measurements (higher price). The cards for dynamic measurements are equipped with the parallel A/D converters as are the signal analysers.

## 2.3   Signal Analysis Based on Fourier Transform

### 2.3.1   Time and Frequency Domain

The primary outcome of measurements of physical quantity which are randomly oscillating in time with a high frequency such as sound pressure and vibration are time records. Inspection of these records provides the first information. Amongst the most important of this information is whether sensors can be measured without error and whether the record is stationary or non-stationary, or contains very high levels of background noise. For slowly varying signals, we can formulate some meaningful conclusions. With the exception of the sinusoidal vibration we will find very little when inspecting the details of the time history of the sound pressure or vibration. If the time history of the signals seems to be random or even chaotic and the

**Table 2.5**   Connectors used in the field of sound and vibration measurements.

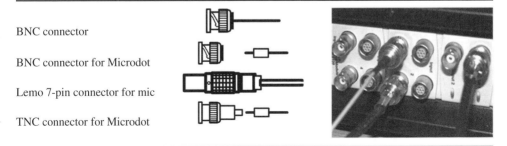

BNC connector

BNC connector for Microdot

Lemo 7-pin connector for mic

TNC connector for Microdot

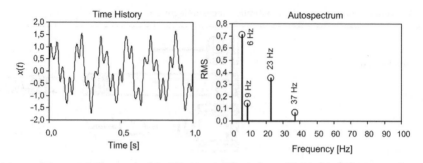

**Figure 2.11**    Presentation of a signal in the time and frequency domain.

inspection fails in the time domain then we should turn our attention to the frequency domain. We compute a frequency spectrum of the sound pressure and vibration.

The advantages of calculating the frequency spectrum in comparison with the time history of a signal is shown in Figure 2.11. The time history of the signal in the left panel seems to be partly random, with a certain regularity in the repetition of the waveform parts. Frequency spectrum (Autospectrum) in the right panel of Figure 2.11 gives information about the frequency and amplitude or RMS of four sinusoidal components of the time signal in the left panel. Frequency analysis allows the origin of the signal components to be revealed.

Vibration and noise spectra produced by machines are composed of tonal components, which are originated by gears, bearings, engines or motors, and so on. The amplitude of the tonal components reflects the technical state of the machine or transmission parts. The spectral analysis is a useful tool for machine diagnostics.

## 2.3.2  Fourier Series for Periodic Functions

The Fourier transform originated in the solution of heat conduction in a solid body which was suggested by Joseph Fourier in 1807. The basic trick consists in using the trigonometric polynomial approximation for solving partial differential equations. This polynomial is now called the Fourier series. For calculation of the coefficients of the Fourier polynomial used orthogonality of trigonometric functions

$$\int_{-\pi}^{+\pi} \cos{(nx)}\,dx = 0, \quad \int_{-\pi}^{+\pi} \sin{(nx)}\,dx = 0, \dots\dots\dots\dots\text{for an integer } n \geq 0$$

$$\int_{-\pi}^{+\pi} \sin{(mx)} \cos{(nx)}\,dx = 0, \quad \dots\dots\dots\dots\dots\dots\dots\text{for any integers } m, n$$

$$\int_{-\pi}^{+\pi} \sin{(mx)} \sin{(nx)}\,dx = 0, \quad \int_{-\pi}^{+\pi} \cos{(mx)} \cos{(nx)}\,dx = 0, \text{for integers } m \neq n$$

$$\int_{-\pi}^{+\pi} \sin^2{(nx)}\,dx = \pi, \quad \int_{-\pi}^{+\pi} \cos^2{(nx)}\,dx = \pi \dots\dots\dots\text{for an integer } n \geq 0. \quad (2.33)$$

Note that the integrated trigonometric functions have an integer number $n$ or $m$ of periods in the integration interval $x \in \langle -\pi, +\pi \rangle$.

For sampled sinusoidal signal segments of the length $N$ exact orthogonality holds only for the harmonics of the sampling frequency $f_S$ divided by $N$. This means that it holds only for the frequencies, which are harmonics (an integer multiple) of the frequency $f_S/N$ (in Hz). Such trigonometric functions are pairwise orthogonal.

Let $x(t)$ be an integrable periodic signal of period $T$, where $t$ is the continuous time. The integrable function is such a function that integral exists. The signal is called periodic if

$$x(t) = x(t + T). \tag{2.34}$$

For a function of time that is integrable on the interval $t \in \langle 0, T \rangle$, the coefficients

$$a_k = \frac{1}{T} \int_0^T x(t) \cos(2\pi kt/T)\, dt, \quad b_k = \frac{1}{T} \int_0^T x(t) \sin(2\pi kt/T)\, dt, \quad k = 0,\ 1,\ 2,\ldots$$

$$\tag{2.35}$$

are called the Fourier coefficients of $x(t)$. The infinite sum

$$\frac{a_0}{2} + \sum_{k=1}^{+\infty} \left[ a_k \cos(2\pi kt/T) + b_k \sin(2\pi kt/T) \right] \tag{2.36}$$

is called the Fourier series of $x(t)$. The Fourier series in engineering applications is generally assumed to converge everywhere except at discontinuities. If the limit of $x(t)$ at some value of time $t$ from right and left differs, then the Fourier series sum at this point of discontinuity is equal to the arithmetic average of the one-sided limits. An example of a sawtooth signal with discontinuities is shown in Figure 2.12. Whether the function is defined in the time instants $T$, $2T$, and so on however, the sum of the Fourier series is the arithmetic average of the one-sided limits.

The exponential Fourier series is the infinity sum of the form

$$x(t) = \sum_{k=-\infty}^{+\infty} F_k \exp\left( j\frac{2\pi}{T}kt \right) \tag{2.37}$$

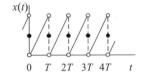

**Figure 2.12** Presentation of a signal in the time and frequency domain.

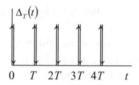

**Figure 2.13** Dirac comb.

where $j$ is an imaginary unit. The Fourier coefficients are given by

$$F_k = \frac{1}{T} \int\limits_0^T x(t) \exp\left(-j\frac{2\pi}{T}kt\right) dt, \quad k = 0, \pm1, \pm2, \dots. \tag{2.38}$$

Derivation of the discrete Fourier transform requires knowing the properties of the Dirac comb function which is shown in Figure 2.13. This function is also known as an impulse train and sampling function. The Dirac comb is the integrable periodic function and it is defined by the formula

$$\Delta_T(t) = \sum_{n=-\infty}^{+\infty} \delta(t - nT). \tag{2.39}$$

The Fourier coefficients $F_k$ for the Dirac comb are

$$F_k = \frac{1}{T} \int\limits_0^T \delta(t) \exp\left(-j\frac{2\pi}{T}kt\right) dt = \frac{1}{T} \exp(-j0) = \frac{1}{T}, \quad k = 0, \pm1, \pm2, \dots. \tag{2.40}$$

All Fourier coefficients are equal to the same value, therefore,

$$\Delta_T(t) = \sum_n \delta(t - nT) = \frac{1}{T} \sum_{k=-\infty}^{+\infty} \exp\left(j\frac{2\pi}{T}kt\right). \tag{2.41}$$

### 2.3.3    Fourier Transform of the Continuous-Time Functions

The result of the Fourier transform serves to calculate the frequency spectrum of the signal. The Fourier transform is defined for continuous time and on the base of this definition the discrete Fourier transform for discrete time is derived. Let $x(t)$ be a complex function of time on the infinite time interval $t \in (-\infty, +\infty)$. The direct and inverse Fourier transform are given by the following formula

$$X(\omega) = \int\limits_{-\infty}^{+\infty} x(t) \exp(-j\omega t) dt \tag{2.42}$$

$$x(t) = \frac{1}{2\pi} \int\limits_{-\infty}^{+\infty} X(\omega) \exp(j\omega t) d\omega. \tag{2.43}$$

**Table 2.6**   Fourier transform of some time functions.

| Signal | Fourier transform | Signal | Fourier transform |
|---|---|---|---|
| $x(t), -\infty < t < +\infty$ | $X(\omega), -\infty < \omega < +\infty$ | $\delta(t)$ | $1$ |
| $ax(t) + by(t)$ | $aX(\omega) + bY(\omega)$ | $1$ | $2\pi\delta(\omega)$ |
| $x'(t)$ | $j\omega X(\omega)$ | $\cos(\omega_0 t)$ | $\pi\left(\delta(\omega - \omega_0) + \delta(\omega + \omega_0)\right)$ |
| $\int x(t)dt$ | $X(\omega)/j\omega + \pi X(0)\,\delta(\omega)$ | $1/\pi t$ | $-j\,sign(\omega)$ |
| $x(t - t_0)$ | $X(\omega)\exp\left(-j\omega t_0\right)$ | $sign(t)$ | $1/j\omega$ |
| $x(t)\exp\left(j\omega_0 t\right)$ | $X(\omega - \omega_0)$ | $x(t)\,y(t)$ | $1/2\pi X(\omega) \otimes Y(\omega)$ |
| $x(at), a \neq 0, a \in R$ | $1/|a|X(\omega/a)$ | $x(t) \otimes y(t)$ | $X(\omega)\,Y(\omega)$ |

Both the direct and inverse transform exist provided that the following integral

$$\int_{-\infty}^{+\infty} |x(t)|\, dt \tag{2.44}$$

exists and there is a finite number of discontinuities.

The Fourier transform maps the time domain to the frequency domain and vice versa. We can say for the direct Fourier transform that exactly one $X(\omega)$ corresponds to each $x(t)$ and any $x(t)$ can be reached from $X(\omega)$.

The Fourier transform not only allows signals in the frequency domain to be presented, but also some calculations in the frequency domain to be performed. This is easier than in the time domain and then the results transform back into the time domain. The transformation formulas are given in Table 2.6. To reach an understanding of some of them firstly we will define some symbols and functions. The symbol $\otimes$ designates convolution. Convolution in the time domain is given by the formula

$$x(t) \otimes y(t) = \int_{-\infty}^{+\infty} x(t - \tau)\,y(\tau)\,d\tau = \int_{-\infty}^{+\infty} x(\tau)\,y(t - \tau)\,d\tau \tag{2.45}$$

while convolution in the frequency domain is defined by formula

$$X(\omega) \otimes Y(\omega) = \int_{-\infty}^{+\infty} X(\omega - \Omega)\,Y(\Omega)\,d\Omega. \tag{2.46}$$

The other functions in Table 2.6 are defined by the following formula

$$sign(t) = \begin{cases} 1/2, & t > 0 \\ -1/2, & t < 0 \end{cases} \qquad X(0) = \int_{-\infty}^{+\infty} x(t)\,dt \tag{2.47}$$

Integration in the time domain corresponds to division of a transformed signal by $j\omega$ in the frequency domain. Due to the function $\delta(\omega)$ this calculation method is unusable for the zero frequency, but in practical terms this limit of integration in the time domain does not restrict the applicability of calculation, for example, conversion of an acceleration signal to the velocity signal.

If the impulse response of the system is specified, that is, the response to the Dirac delta function at the input of the system, then the convolution of the input signal $x(t)$ and the impulse response $y(t)$ is the output signal of the system.

The Fourier transform preserves the energy of signals, which is defined as a time integral of the square of the absolute value of a complex time function $\|...\| = \int_{-\infty}^{+\infty} |...|^2 \, dt$. This property of the Fourier transform is known as Parseval's theorem. Without detailed proof we can say that the following equation is valid

$$\|x(t)\| = \frac{1}{2\pi} \int_{-\infty}^{+\infty} |X(\omega)|^2 \, dt. \tag{2.48}$$

The validity of the previous equation can be proved using the Fourier transform of the convolution of a function $x(t)$ and the complex conjugate function $y(t) = x(t)^*$ in the time domain.

The Fourier transform has a localisation property which means that the more concentrated a function of time, the more spread the Fourier transform has as a function of frequency. This property is demonstrated on the Gaussian function in Figure 2.14. If the signal in the time domain becomes narrower and narrower and tends to approach the Dirac delta function then the signal in the frequency domain extends to infinity.

## 2.3.4 Short-Time Fourier Transform

The short-time Fourier transform (STFT) differs from the general Fourier transform in its use of a time window $w(t)$, which is nonzero for only a short period of time and zero everywhere else. The time window position with respect to time determines the parameter $\tau$. The result

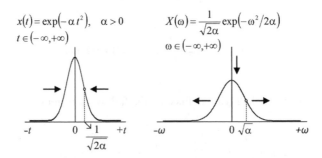

**Figure 2.14**  Localisation property of the Fourier transform.

of STFT calculation does not depend only on the angular frequency $\omega$, but also on the parameter $\tau$

$$X(\omega, \tau) = \int_{-\infty}^{+\infty} x(t)\, w(t - \tau) \exp(-j\omega t)\, dt \tag{2.49}$$

The rectangular or Hanning window is frequently used for STFT calculation. The group of the STFT spectra which differs in time $\tau$ is also called multispektrum.

### 2.3.5 Fourier Transform of the Discrete-Time Functions

The Fourier transform of functions with continuous time was defined on the time interval from minus infinity to plus infinity. Records of digitised signals always contain a finite number of samples. For further calculations, we will assume that this record is a period of an infinite sequence of samples. We believe that this record represents the signal properly, why not?

Let $x_n$ be a period of a discrete signal of the finite time duration $T = NT_S$

$$x_n = x_{n+N}, \quad n = 0, \pm 1, \pm 2, \dots. \tag{2.50}$$

The bridge between the analogue and digital world is built by the formula $x(t) = \sum_n x_n \delta(t - nT_S)$. As it is assumed that the infinity sequence of samples is periodic, the continuous time signal can be defined with the use of one period of this sequence

$$x(t) = \sum_{n=0}^{N-1} \sum_k x_n \delta\big(t - (n + kN) T_S\big). \tag{2.51}$$

The Fourier transform of the continuous time signal is as follows

$$X(\omega) = \int_{-\infty}^{+\infty} \sum_{n=0}^{N-1} \sum_k x_n \delta\big(t - (n + kN) T_S\big) \exp(-j\omega t)\, dt$$

$$= \sum_{n=0}^{N-1} x_n \exp\big(-j\omega n T_S\big) \sum_k \exp(-j\omega k T). \tag{2.52}$$

The infinite sum of exponential functions in the previous formula can be replaced by the formula for calculation of the Dirac comb (2.41) with substitutions $\omega$ for $t$ and $2\pi/T$ for $T$. We obtain

$$X(\omega) = \sum_{n=0}^{N-1} x_n \exp\big(-j\omega n T_S\big)\, 2\pi/T \sum_k \delta(\omega - k2\pi/T)$$

$$= \sum_{n=0}^{N-1} x_n \exp(-j2\pi nk/N)\, 2\pi/T \sum_k \delta(\omega - k2\pi/T). \tag{2.53}$$

The Fourier transform as a function of the continuous angular frequency $\omega$ contains a Discrete Fourier transform which deals only with the finite sequence of samples. Nonzero values of $X(\omega)$ are only for $\omega = k2\pi/T$ where $k = 0, 1, 2, \ldots$ Substituting $X_k$ for $\sum_{n=0}^{N-1} x_n \exp(-j2\pi nk/N)$ in the Eq. (2.53) results in the formula

$$X(k2\pi/T) = X_k 2\pi/T\delta(0). \tag{2.54}$$

Due to the periodicity exponential function the theoretically different result of calculation of $X_k$ is only for $k = 0, 1, 2, \ldots, N - 1$.

The direct Discrete Fourier transform (DFT) is given by the formula

$$X_k = \sum_{n=0}^{N-1} x_n \exp(-j2\pi nk/N), \quad k = 0, 1, \ldots, N - 1. \tag{2.55}$$

Due to the number of the input samples, this transformation is also known as the $N$-point DFT. The formula for calculation of the inverse DFT is given as follows

$$x_n = \frac{1}{N} \sum_{k=0}^{N-1} X_k \exp(j2\pi nk/N), \quad n = 0, 1, \ldots, N - 1 \tag{2.56}$$

The Discrete Fourier transform of $N$ input real or complex samples again produces $N$ output values which are complex.

## 2.3.6   Inverse Fourier Transform of the Discrete-Time Functions

The formula for the calculation of DFT was derived in the previous chapters. In this subsection the formula for calculating the inverse DFT shall be verified. The formulas (2.55) and (2.56) have the form of a product of a matrix and a vector. Samples of the signal can be arranged in a column vector $\mathbf{x} = [x_0, x_1, \ldots, x_{N-1}]^T$ (superscript $T$ denotes the transpose of the vector) and the calculated data in another column vector $\mathbf{X} = [X_0, X_1, \ldots, X_{N-1}]^T$ of the same size. The values of individual samples are multiplied by a complex factor

$$W_N^{nk} = \exp(-j2\pi nk/N) = \cos(j2\pi nk/N) - j\sin(j2\pi nk/N) \tag{2.57}$$

which is called a twiddle factor. All the twiddle factors can be arranged into an $N$ by $N$ symmetric matrix $\mathbf{W} = \left[W_N^{nk}\right]$ where $n$ is the index of the row and $k$ is the index of the column of this matrix

$$\mathbf{W} = \begin{bmatrix} W_N^{0\times 0} & W_N^{0\times 1} & \cdots & W_N^{0\times(N-1)} \\ W_N^{1\times 0} & W_N^{1\times 1} & \cdots & W_N^{1\times(N-1)} \\ \vdots & \vdots & \ddots & \vdots \\ W_N^{(N-1)\times 0} & W_N^{(N-1)\times 1} & \cdots & W_N^{(N-1)\times(N-1)} \end{bmatrix}. \tag{2.58}$$

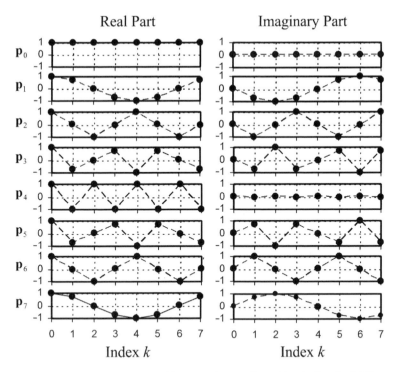

**Figure 2.15** Complex sinusoid $\mathbf{p}_n, n = 0, 1, ..., 7$ used by the 8-point DFT.

With the new designation which is based on the matrix and vectors the direct Discrete Fourier transform (DFT) may be written in matrix form as follows

$$\mathbf{X} = \mathbf{W}\mathbf{x}. \tag{2.59}$$

The absolute value of this factor $W_N^{kn}$ is equal to unity. The $n$-th row of the matrix $\mathbf{W}$ can be designated by $\mathbf{p}_n = \left[ W_N^{n \times 0}, W_N^{n \times 1}, ..., W_N^{n \times (N-1)} \right]$. The real part of the row vector $\mathbf{p}_n$ contains the $n$ periods of the cosine function while the imaginary part of the same row vector contains the sine function. The complex sinusoid $\mathbf{p}_n, n = 0, 1, ..., 7$ used by DFT for $N = 8$ is shown in Figure 2.15.

The $k$-th column of the matrix $\mathbf{W}$ can be designated by $\mathbf{q}_k^T = \left[ W_N^{0 \times k}, W_N^{1 \times k}, ..., W_N^{(N-1) \times k} \right]^T$. The frequency of the mentioned cosine and sine functions depends on the row or column index. The number of periods is equal to a whole number. As mentioned in Section 2.3.2 a set of rows or columns are pairwise orthogonal, therefore such different nonzero vectors are linearly independent. Due to the orthogonality of the different rows the determinant of the matrix $\mathbf{W}$ is nonzero and the matrix is invertible. The different columns of the matrix $\mathbf{W}$ are also orthogonal. For a given vector $\mathbf{X}$ it is possible to compute the unknown vector $\mathbf{x}$

$$\mathbf{x} = \mathbf{W}^{-1}\mathbf{X}. \tag{2.60}$$

The inverse matrix of the matrix $\mathbf{W}$ is given by the formula

$$\mathbf{W}^{-1} = \frac{1}{N}\mathbf{W}^* \tag{2.61}$$

The matrix product of the matrix $\mathbf{W}$ and the corresponding inverse matrix has to be equal to the identity matrix $\mathbf{W}\mathbf{W}^{-1}/N = \mathbf{E}$. The corresponding elements of both matrices are mutually complex conjugate. The different rows $\mathbf{p}_n^*$ of the complex conjugate matrix $\mathbf{W}^*$ are linearly independent like the different rows of the matrix $\mathbf{W}$. The same rule applies to the columns $\mathbf{q}_k^*$ of this matrix. Due to the orthogonality and with respect to the formula (2.57) and the identity $W_N^{nk}\left(W_N^{kn}\right)^* = 1$ we obtain

$$\mathbf{p}_n\mathbf{q}_k^* = \begin{cases} 0, & n \neq k \\ N, & n = k. \end{cases} \tag{2.62}$$

The formula for calculation of the inverse DFT has been verified.

### 2.3.7  DFT of a Constant and a Cosine Signal

Let $x_n = a$ be a finite sequence of constant samples where $n = 0, 1, 2, \ldots, N - 1$. The value of $X_0$ is equal to $N$

$$X_0 = \sum_{n=0}^{N-1} x_n \exp\left(-j2\pi nk/N\right) = a \sum_{n=0}^{N-1} \exp\left(0\right) = N \tag{2.63}$$

For $k > 0$ we obtain

$$X_k = \sum_{n=0}^{N-1} x_n \exp\left(-j2\pi nk/N\right) = a \sum_{n=0}^{N-1} W_N^k \tag{2.64}$$

where $W_N^k$ is the previously defined twiddle factor. The twiddle factor can be considered as a vector in the complex plane. If we add these vectors in such a way that the initial point of the vector is placed at the terminal point of the previous vectors, then all the twiddle factors form an equiangular and cyclic polygon whose vertices lie on a circle in the complex plane as shown in Figure 2.16. The polygon is closed; the sum of the corresponding vectors is equal to zero.

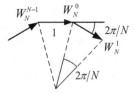

**Figure 2.16**  Twiddle factors in the form of an equiangular cyclic polygon.

Notice the fact that $X_0$ is equal to the product of the sample value and the sample number while $X_k$ is equal to zero for any $k > 0$.

The DFT of a cosine signal depends on the number of waves per the record which is to be transformed. Firstly, we calculate the discrete Fourier transform of a cosine signal with the integer number of waves (whole number of periods in $N$ samples) using an analytical approach. We assume that the signal samples are given by the formula

$$x_n = A \cos (2\pi Mn/N + \varphi), n = 0, 1, \ldots, N - 1 \qquad (2.65)$$

where $M$ is an integer number $0 < M \le N/2$. $M$ is equal to the number of waves per record of the length $N$. The advantage of the integer number of waves is that at the connecting points of the two adjacent records the periodically extended signal is smooth, without any jump.

Using the formula $\cos (x) = (\exp (jx) + \exp (-jx))/2$ the cosine function is changed into the sum of the exponential functions

$$x_n = A \exp (j (2\pi Mn/N + \varphi)) + \exp (-j (2\pi Mn/N + \varphi)) / 2, \quad n = 0, 1, \ldots, N - 1 \qquad (2.66)$$

After substituting the samples $x_n, n = 0, 1, \ldots, N - 1$ for Eq. (2.55) in the discrete Fourier transform formula

$$X_k = \frac{A}{2} \sum_{n=0}^{N-1} \left[ \exp (j (2\pi Mn/N + \varphi)) + \exp (-j (2\pi Mn/N + \varphi)) \right] \exp (-j2\pi nk/N)$$

$$= \frac{A}{2} \sum_{n=0}^{N-1} \left[ \exp (j (2\pi (M - k) n/N + \varphi)) + \exp (-j (2\pi (M + k) n/N + \varphi)) \right]. \qquad (2.67)$$

we obtain

$$X_M = A \frac{N}{2} \exp (j\varphi), \quad X_{N-M} = A \frac{N}{2} \exp (-j\varphi), \quad X_k = 0, k \ne M, N - M \qquad (2.68)$$

Notice the fact that the absolute value of $X_M$ and $X_{N-M}$ is equal to the product of half the number of samples $N/2$ and the cosine signal amplitude while $X_k$ is equal to zero for arbitrary $k \ne M$ and $k \ne N - M$ as shown in Figure 2.17. In other words, we can deduce that the only components of the output vector with the indexes $M$ and $N - M$ are nonzero, while the others

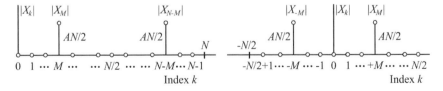

**Figure 2.17** Two-sided DFT spectrum with two ways of indexing ($N$ is an even number).

are equal to zero. Also note that the indexes of the nonzero components are symmetric or about half the number $N$ and that these components are complex conjugate

$$X_M = X^*_{N-M}. \tag{2.69}$$

The complex conjugate of the result of calculation of DFT is valid only for real signals. It is possible to prove that this property is not valid for complex signals.

As one can see, there is a complex conjugate symmetry in the DFT around a component with the index of $N/2$. The complex conjugate components in ascending order, starting with the index 1, are complex conjugate with the components in descending order, starting from the index $N-1$. It is also possible that the indices $N-1, N-2, \ldots$ can be replaced by indices $-1, -2, \ldots$, then the complex conjugate symmetry will be around the component with index 0 instead of $N/2$ as shown on the right side in Figure 2.17. Negative indexes justify the use of the term negative frequencies.

### 2.3.8 Phasors as a Tool for Modelling Harmonic Signals

The model of a harmonic signal can be created using a phasor, which is the position vector in the complex plane with the origin at zero, while the endpoint rotates along the circle at a constant rotational speed. Projection of the phasor into the real axis results in a vector of variable length. The vector length as a sinusoidal function of time creates a harmonic signal.

The projection may be replaced by the sum of the two phasors that rotate against each other. A harmonic signal with amplitude $A$, initial phase $\varphi$ and angular frequency $\omega$ can be modelled as a sum of two phasors of the same length and opposite initial phases as shown in Figure 2.18. Both the phasors are rotating at the same angular velocity, but in an opposite direction

$$X^+ = \frac{A}{2}\exp\left(+j\left(\omega t + \varphi\right)\right), \quad X^- = \frac{A}{2}\exp\left(-j\left(\omega t + \varphi\right)\right) \Rightarrow X^+ = (X^-)^*. \tag{2.70}$$

At any time the sum of both the phasors is a vector of the direction of the real axis and the length of this vector changes periodically in time. The phasors in the initial position, that is, for the time $t = 0$, are complex conjugate as components resulting from the calculation of the

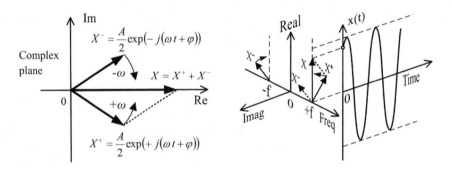

**Figure 2.18**  Model of a real harmonic signal based on the sum of two phasors.

Fourier transform. The diagram on the right side of Figure 2.18 is produced if the diagram on the left side of this figure is expanded by the frequency axis. This shows the direction and the angular velocity or rotational frequency of the two phasors. Given that the components resulting from the calculation of DFT can be indexed in the following way

$$-N/2+1, ..., -2, -1, 0, +1, +2, ..., +N/2,$$

the 3D diagram illustrates the relationship between the DFT and the model of sinusoidal signals based on the rotating phasors. The length of the phasor corresponds to half the amplitude of a sinusoidal component. The result of calculation of DFT according to the formulas differs from phasors only in the absolute value of the corresponding complex number resulting from the calculation of DFT. This value is $N/2$ times greater than the length of the phasor. The difference consists only in the scale factor as the phase of the sinusoidal components is the same as the angle of the complex number. The advantage of a 3D diagram is also that the angle of the DFT components in the complex plane, as well as the initial phase of the rotating phasors, is displayed in an illustrative manner.

The effect of phasor rotating at different speeds is shown in Figure 2.19. Three odd harmonics of the fundamental frequency generate a periodic signal that approaches a rectangular signal of 50% duty cycle.

If both phasors differ in their length and the initial phases, while rotating against each other at the same speed, then the sum of them is not a real number, but a complex. The signal which is corresponding to these phasors is complex, that is, this signal is composed of the real and imaginary part. The endpoint of the phasor sum $X^+ + X^-$ creates an ellipse called an orbit as shown in Figure 2.20.

It is assumed that time is continuous and therefore phasor rotation is also continuous in time and the endpoint of a phasor moves around the circle. If the time is discrete then the rotation is not continuous and it consists of jumps between isolated points on the circumference of a circle. The rotational angle is increasing by increments of the same angle difference. This

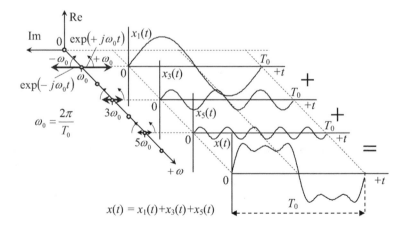

**Figure 2.19**  Model of a periodic signal based on three pairs of phasors rotating at different speeds.

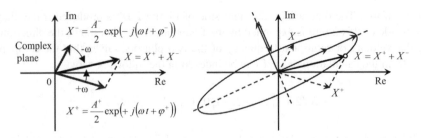

**Figure 2.20**   Model of a complex harmonic signal based on the sum of two phasors.

property does not restrict the use of the phasors. The phasors can be used if it is possible to decompose the signal to harmonic (sinusoidal) components.

### 2.3.9  Example of DFT Calculation

The purpose of the example is to compare the result of the DFT calculation of a record containing an integer and a non-integer number of waves. Let us take for example a signal with three and three and a half waves. The number of samples is equal to 16. Note also that the numbers. 16 and 3 are mutually prime. The DFT of the record with three waves is shown in Figure 2.21 and the DFT of the record containing three and a half waves is shown in Figure 2.22. In the first case the diagram contains the real and imaginary part while the magnitude of the complex numbers is shown in the second case. As was mentioned before the record with three waves can be smoothly connected and periodically expanded, while in the second case, there is a discontinuity at the junction of the adjacent records.

The calculation of DFT for the sinusoidal input signal with three waves in the 16 samples results in zero everywhere, but at the indexes which are equal to 3 and $16 - 3 = 13$ the spectrum amplitude is equal exactly to the value which is a product of the sinusoidal signal amplitude and half the number of samples. The frequency of this signal with 3.5 waves corresponds

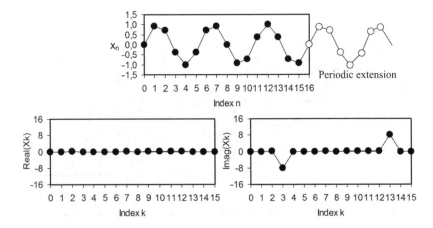

**Figure 2.21**   DFT of a cosine signal with 3 waves in a sequence of 16 samples.

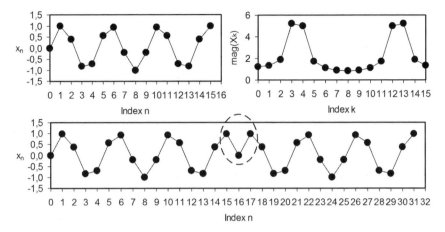

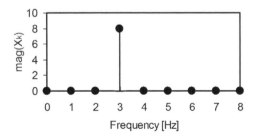

**Figure 2.22**   DFT of a cosine signal with 3.5 waves in a sequence of 16 samples.

neither to the index 3 nor to the index 4. The amplitude of the spectrum components of the indices 3 and 4 is also less than the value of 8, which corresponds to the value for the integer number of waves in the 16 samples.

## 2.3.10   Time and Frequency Scales

The index $n$ is on the horizontal axis of the time domain signal and the index $k$ is on this axis for DFT. The time scale can be used instead of the index $n$ and the frequency scale can replace the index $k$ by time or frequency

$$
\begin{aligned}
t_n &= nT_S, \dots\dots\dots\dots\dots\dots\ n = 0, 1, 2, \dots, N - 1, \\
f_k &= k/NT_S = k/T, \dots\dots k = 0, 1, 2, \dots, N - 1,
\end{aligned}
\tag{2.71}
$$

where $T_S$ is the sampling interval, $N$ is the sample number for calculation of DFT and $T = NT_S$ corresponds to the length of the time record of $N$ samples at the sampling interval of $T_S$. Due to the symmetry of the two-sided spectrum a one-sided spectrum is used for the real signals. An example for a signal from Figure 2.21 is shown in Figure 2.23.

**Figure 2.23**   One-side DFT spectrum (sampling interval 1/16 s).

Assuming that the number of samples is even the frequency range of a one-sided spectrum is restricted to half the sampling frequency. If the sample number $N$ is an even number then the index $k$ for calculation frequency is from the interval of $k = 0, 1, 2, ..., N/2$.

When considering the transition band of the antialiasing filter and the number of the sample is a power of two then the frequency range is limited by even more than half the sampling frequency as explained above. The index $k$ for calculation frequency is from the interval of $k = 0, 1, 2, ..., N/2.56$. The signal analysers which are supplied by the Brüel & Kjær Company use the sampling rate, which is a power of two in hertz (Hz). The frequency ranges of measurements for the sampling frequency equal to a power of two are then limited by the smart values, such as 100, 200, 400, 800 Hz, and so on. These frequency ranges form a geometric sequence with a scale factor which is equal to 2. Other sequences include the frequency ranges such as the range of 1000 Hz.

### 2.3.11   Spectral Unit of Autospectrum

The frequency spectrum which is contained within the calculated absolute value of DFT is not the one used for practice because it relies on the length of the record for calculation of DFT. It is more suitable to scale the vertical axis of the frequency spectrum in units as root mean square (RMS), power (PWR) or power spectral density (PSD), that is, power related to the frequency band of 1 Hz. The spectral units or scales are listed in Table 2.7.

The calculation of the spectrum for a constant signal corresponding to the zero frequency (index $k = 0$) indicates that the result of the DFT is a value which is equal to the product of this constant and the number of samples. If the number of waves per record of the length $N$ is equal to index $k$ then the absolute value of the calculation of DFT for such a sinusoidal signal is equal to the product of the signal amplitude and half the number of sample. The vertical axis of the spectrum depends on the length of the record $N$ for calculating the DFT. The diagnostics of machines therefore prefers the RMS of the sinusoidal components of the frequency spectrum

$$RMS_k = \begin{cases} |X_0|/N, & k = 0, \\ |X_k|/\left(\sqrt{2}N/2\right), & k = 1, 2, ..., N/2 \end{cases} \tag{2.72}$$

where $X_k$ is given by the formula (2.55). The units on the vertical axis of the RMS spectrum are identical with the units of the signal, such as Pa, m/s, m/s$^2$, and so on.

The scale of the vertical axis in the power (PWR) is mainly used for acoustic signals, but PWR can also be used for other signals. The units of the PWR spectrum are the squared unit

**Table 2.7**   Spectral unit of autospectrum.

| Scale | Name | Used for signals | Formula |
|-------|------|------------------|---------|
| RMS | Root Mean Square | Diagnostics of machines | $RMS = \text{amplitude}/\sqrt{2}$ |
| PWR | Power | Acoustics | $PWR = RMS^2$ |
| PSD | Power spectral density | Random process | $PSD = PWR/\Delta f$ |

**Table 2.8** Reference values for calculation of decibels.

|  | Sound pressure | Velocity | Acceleration | Force | Voltage |
|---|---|---|---|---|---|
| *ref* | $2 \times 10^{-5}$ Pa | $1 \times 10^{-6}$ mm/s | $1 \times 10^{-6}$ m/s$^2$ | $1 \times 10^{-6}$ N | 1 V |

of the corresponding signal, such as Pa$^2$, (m/s)$^2$, (m/s2)$^2$, and so on. The relationship between the PWR and RMS scale is given by the formula

$$PWR_k = RMS_k^2, \quad k = 0, 1, 2, \ldots, N/2 \qquad (2.73)$$

The power spectral density (PSD) is used for random signals. The formula for calculating the PSD spectrum is as follows

$$PSD_k = PWR_k/\Delta f, \quad k = 0, 1, 2, \ldots, N/2 \qquad (2.74)$$

where $\Delta f = 1/T = 1/(NT) = f_S/N$ is a distance in hertz (Hz) between adjacent spectrum components.

The use of decibels reduces the numerical range of very large or small numbers. The conversion of the RMS of a physical quantity, denoted by $y$, to dimensionless decibel is carried out according to the following formula

$$dB/ref = 20 \log \left( \frac{y}{ref} \right) \qquad (2.75)$$

where *ref* is a reference value given in Table 2.8.

The decibels are widely used as a measure of increase or decrease of the signal power or RMS. The decibels are in this book mainly for evaluating the power or intensity of sound. Assessment of sound or acoustic pressure in the physical unit of pressure with the use of Pa (Pascal) or $\mu$Pa, is replaced by the conversion of the absolute value to the decibels. Similarly, the decibels can be used for assessment of vibration. If any physical quantity is expressed in decibels, it means that it is designated as a level of this physical quantity, for example a sound pressure level (SPL).

### 2.3.12   Cross-Spectrum

Vibration power produced or absorbed at a certain frequency can also be calculated from the two signals such as vibrations in units of speed and dynamic force in Newtons. Vibration power which is generated by a machine is measured. When deriving formulas for calculation it should differentiate between vectors and phasors.

Because we are interested in the physical quantities as sinusoidal functions of time the calculation of mechanical power necessitates taking into account the different possible phases between force and velocity. A method of converting the phasor of acceleration or displacement to the phasor of velocity is shown in Figure 2.24. The index $k$ designates a component in the Fourier spectrum. The velocity vector is calculated by integrating acceleration or by calculation

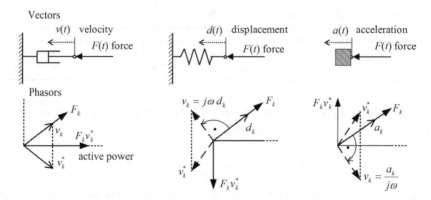

**Figure 2.24**  Caculation of the mechanical power.

of the first derivative of the displacement with respect to time. The phasor of displacement or velocity needs to rotate by 90 degrees in relation to the phasor of velocity. As the electrical power can be real or active and reactive, so it is the same in mechanics. The mechanical power is the scalar product of force and velocity vectors while when using the phasors in the complex plane the power is a product of two complex numbers which corresponds to the phasors in their initial positions. The resulting complex number has the real and imaginary part, that is, the active and reactive power.

Now, we will deal with two different signals $x(t)$, $y(t)$ with the corresponding phasors $X_k$, $Y_k$ whose speed of rotation determines the index $k$. The procedure to calculate the same variable as power with the use of two signals is shown on the right in Figure 2.25. On the left in this figure the way to calculate the signal power which was derived earlier is shown. If both signals are different, then the calculation procedure for mechanical power results in a cross-spectrum designated shortly by CROSS. This spectrum is complex. If the input signals correspond to the dynamic force and vibrations in velocity then the CROSS spectrum corresponds to the mechanical power. The formula for calculating the power of a signal and cross spectra of two signals is as follows:

$$PWR_k = \begin{cases} (X_0/N)^2, & k = 0, \\ 2X_kX_k^*/N^2, & k = 1, 2, ..., N/2 - 1. \end{cases} \quad (2.76)$$

$$CROSS_k = \begin{cases} T X_0 Y_0^*/N^2, & k = 0 \\ 2 T X_k Y_k^*/N^2, & k = 1, 2, ..., N/2 - 1 \end{cases} \quad (2.77)$$

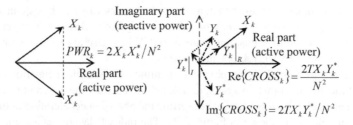

**Figure 2.25**  Physical meaning of a component of the power spectrum and the cross-spectrum.

where $T = 1/f_1 = NT_S = N/f_S$ is the length of the time record of $N$ samples for calculation of DFT and $k = Nf_k/f_S$ is an index.

### 2.3.13 Calculation of the Inverse Discrete Fourier Transform

The formula for calculation of the discrete inverse Fourier transform is as follows

$$x_n = \frac{1}{N} \sum_{k=0}^{N-1} X_k \exp(j2\pi nk/N), \quad n = 0, 1, ..., N-1. \tag{2.78}$$

The formula for calculating the inverse DFT is similar in structure to the formula for calculating the direct DFT. We solve the problem of how to calculate the direct and inverse DFT using a unified algorithm. If the complex conjugate operation is applied two times then because $x_n = \left(x_n^*\right)^*$ this yields

$$x_n = \left(x_n^*\right)^* = \left( \sum_{k=0}^{N-1} \left( \frac{1}{N} X_k \exp(j2\pi nk/N) \right)^* \right)^*$$

$$= \left( \sum_{k=0}^{N-1} \frac{X_k^*}{N} \exp(-j2\pi nk/N) \right)^*, \quad n = 0, 1, ..., N-1. \tag{2.79}$$

The inverse discrete Fourier transform may be written in the form

$$x_n = \left( \mathrm{DFT} \left\{ \frac{X_k^*}{N} \right\} \right)^*, \quad n = 0, 1, ..., N-1. \tag{2.80}$$

where DFT designates the algorithm for the direct discrete Fourier transform. Input data for the algorithm of the inverse DFT are modified so that the input data for the direct DFC are equal to $X_k^*/N$. Notice the fact that the inverse Fourier transform does not require a special algorithm.

### 2.3.14 Fast Fourier Transform

Because this book is not about the calculation algorithms of fast Fourier transform, it is only appropriate to give a brief historical overview. The Discrete Fourier Transform, where $N$ is an arbitrary number, requires $N^2$ complex multiplications and additions. In addition, the discrete Fourier transform (DFT) requires extensive computational time to evaluate. Techniques for reducing the computational time were first discovered by C. Runge in 1903. Runge essentially described the Fast Fourier Transform (see below). Danielson and Lanczos in 1942 recognised certain symmetries and periodicities which reduced the number of operations. Even with the progress in the field of the digital computer the algorithm for reducing the computational time was unknown until 1965 when James W. Cooley and John W. Tukey published their discovery which has become known as the Fast Fourier Transform (FFT) [1,2]. The fast Fourier transform belongs to the top 10 algorithms with the greatest influence on the development and practice of science and engineering in the twentieth century.

The algorithm solution submitted by Cooley and Tukey assumes that the number of samples in a sequence to be transformed is a power of two $N = 2^m$. For this assumption computational methods for the Fast Fourier Transform are based on two approaches:

- DIT – Decimation in time
- DIF – Decimation in frequency.

Each stage of the FFT algorithm is divided into two algorithms applied to half the number of points. The number of the complex multiplications and additions is equal to $(N/2)^2 + (N/2)^2 = N^2/2$. The effect of the division of calculation into two parts each for half the number of samples is half the number of the mentioned mathematical calculations in total. With respect to the initial number of samples $N = 2^m$, division of the number of samples in halves at each stage of the calculation may be repeated $m$ times. The Cooley-Tukey algorithm for the Fast Fourier Transform requires only $Nm = N \log_2 N$ complex multiplications and additions. It is said that the algorithm is performed in $O\left(N \log_2 N\right)$ time.

For example, if $N = 1024 (= 2^{10})$, then the DFT algorithm requires more than one million mathematical operations while the FFT algorithm only 10 240 operations, which results in a considerable saving of computation time.

The development of an algorithm for the calculation of DFT continued. In 1998 Matteo Frigo and Steven G. Johnson proposed a new solution called the Fastest Fourier Transform in the West. This algorithm calculates the DFT for any number of samples.

The FFTW algorithm is for the record length that has a power-of-two. Comparison of the relative computational speeds for a large range of $\log_2(N)$ related to speed for $N = 256$ is shown in the left panel of Figure 2.26. The FFTW algorithm is designed for arbitrary record length including non-power-of-two. The relative computational speeds for all the possible lengths in between $N = 512$ and 1024 are shown in the right panel of Figure 2.26.

## 2.3.15   Time Window

Examples, which are described in Section 2.3.6, show the dependence of the result of the DFT calculation on the number of waves per record of the input signal. If the number of waves is equal to an integer then it is possible to determine the amplitude of the sinusoidal signal

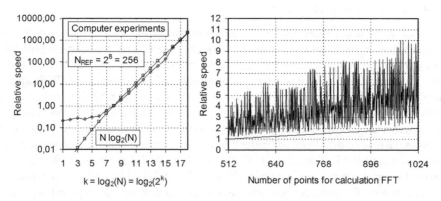

**Figure 2.26**   Comparison of computational speed of the FFT and FFTW algorithm.

without distortion. Frequency of the input sinusoidal signal corresponds to one of the discrete frequencies of the spectrum. If the frequency falls in between the discrete frequencies of the spectrum then the amplitude of the components closest to this frequency is distorted. However, we use the DFT to analyse arbitrary signals from machines. Because we often do not know the exact value of the frequency of the signal components, we will show how to deal with the amplitude distortion.

The orthogonality property of DFT coefficients vector and sinusoidal functions

$$x_n = A \cos \left( 2\pi k \frac{n}{N} \right) = \frac{A}{2} \left( \exp \left( j2\pi k \frac{n}{N} \right) + \exp \left( -j2\pi k \frac{n}{N} \right) \right), \quad n = 0, 1, ..., N - 1. \quad (2.81)$$

determines the result of DFT calculation given by the formula (2.55). If the frequency of this signal matches any discrete frequency of the spectrum (i.e. the index $k$ is a whole number) then the result will be as described in previous chapters. If the frequency of a sinusoidal signal differs from all the discrete frequencies of the spectrum, then the spectrum is smeared and contains a number of nonzero components as shown in Figure 2.22. It is said about this phenomenon, called as a spectral leakage, that the signal energy is divided into frequency bins as a storage place for the spectral energy. The frequency index $k$ is called the bin number, and $|X_k|^2$ can be regarded as the total energy in the $k$-th bin of the range from $k - 0.5$ to $k + 0.5$.

The formula for calculating the DFT assumes that the finite sequence of samples corresponds to one period of an infinitely long lasting signal. We will analyse the effect of shortening the measuring time in a finite interval of the length $T$. The Fourier transform of the continuous harmonic signal $x(t) = A \cos(\omega t + \varphi)$ is as follows

$$X(\omega) = \int_{-\infty}^{+\infty} A \cos \left( \omega_0 t \right) \exp \left( -j\omega t \right) dt = A\pi \left( \delta \left( \omega - \omega_0 \right) + \delta \left( \omega + \omega_0 \right) \right) \quad (2.82)$$

where $\delta(\omega)$ is the Dirac delta function of the angular frequency $\omega$.

Firstly, we introduce a function of continuous time which is known as the rectangular time window $w(t)$. This function is equal to unity inside the interval of length $T$ and zero elsewhere

$$w(t) = \begin{cases} 1, & t \in \langle -T/2, +T/2 \rangle \\ 0, & t \notin \langle -T/2, +T/2 \rangle \end{cases} \quad (2.83)$$

The Fourier transform of the rectangular time window $w(t)$ is as follows

$$W_T(\omega) = \int_{-T/2}^{+T/2} \exp \left( -j\omega t \right) dt = T \frac{\sin(\omega T/2)}{\omega T/2} = T \operatorname{Sinc}(\omega T/2) \quad (2.84)$$

We view the signal $x(t)$ through the window $w(t)$ if the signal is multiplied by the window function. The Fourier transforms of the windowed function is given by the convolution of the

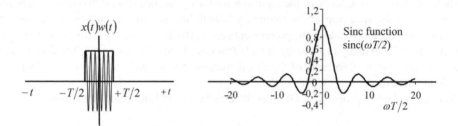

**Figure 2.27**   Windowed sinusoidal function and sinc function.

Dirac delta function (2.82) and the Sinc function (2.84) in the frequency domain

$$X_w(\omega) = \int_{-\infty}^{+\infty} x(t) \, w(t) \exp(-j\omega t) \, dt = \frac{1}{2\pi} X(\omega) * W_T(\omega)$$

$$= A/2 \left( W_T(\omega - \omega_0) + W_T(\omega + \omega_0) \right)$$

$$= TA/2 \left( \mathrm{Sinc} \left( (\omega - \omega_0) \, T/2 \right) + \mathrm{Sinc} \left( (\omega + \omega_0) \, T/2 \right) \right). \tag{2.85}$$

The windowed sinusoidal function and the Sinc function for $\omega_0 = 0$ is shown in Figure 2.27. Convolution of the Sinc function with the Dirac delta function results in a shift by $\pm\omega_0$ in the direction of the axis $\omega$.

In the case of the positive value of the angular frequency $\omega_0$ the Sinc function is shifted by this frequency in the direction of the horizontal axis, as shown in the upper panel in Figure 2.28.

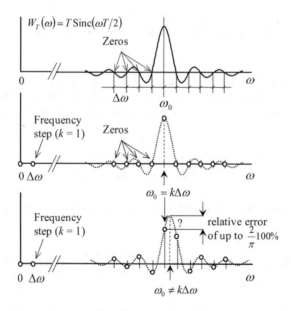

**Figure 2.28**   Effect of the cosine signal frequency on the the magnitude of the spectrum components which are calculated according to (2.85).

For the time interval of length $T$ corresponding to the length of one period for the calculation of DFT the frequency resolution in rad/s of the frequency spectrum is given by the formula $\Delta\omega = 2\pi/T$. The frequency resolution in hertz (Hz) is given by $\Delta f = 1/T$. Let the shift $\omega_0$ of the Sinc function be given by a multiple of the frequency resolution of the spectrum

$$\omega_0 = k\Delta\omega \qquad (2.86)$$

where $k$ is an integer number.

The position of the maxima of the Sinc function in relation to the discrete frequencies of the spectrum determines how much energy falls on each bin of the spectrum. If the frequency $\omega_0$ coincides with some of the discrete frequencies of the spectrum which is calculated according to the formula (2.85), then this relative position is shown in the middle panel in Figure 2.28. If the frequency $\omega_0$ falls in between the mentioned discrete frequencies then this relative position is shown in the low panel of the mentioned figure. The rectangular window is suitable only for signals composed from harmonics of the $\Delta f = 1/T$ frequency.

There are time windows that partially compensate for the errors in the reading of the amplitude of the spectrum components. An overview of the most used time windows is shown in Table 2.9.

The effect of choice of time window is shown in Table 2.10. Input signals for frequency analysis were sampled with 256 Hz and the recording contained 256 samples, that is, the length of the time recording is one second. The frequency spectrum has a vertical scale in RMS. The interval between the adjacent frequencies of the spectrum is equal to 1 Hz. The DFT spectrum was calculated for a sinusoidal signal of the frequency 50 and 50.5 Hz. In the first case the signal frequency falls in a range of the discrete frequencies while in the second case it falls between two adjacent frequencies. Because the amplitude of the sinusoidal signal is equal to unity, the RMS of this signal is equal to $1/\sqrt{2} \approx 0.707$. From Table 2.10 it is clear how large errors when reading RMS can arise depending on the choice of the time window.

The rectangular time window is used only for signals which contain harmonics corresponding to the first nonzero frequency which is equal to $1/T$ where $T$ is the length of the input record. The time window Hanning or Kaiser-Bessel is used if we do not know anything about the frequency composition of the analysed signal. The Flat Top time window is used to calibrate sensors as piezoelectric accelerometers or microphones using the AC signal. Description of the time window by Flat Top is clear from Figure 2.29. The top part of the spectrum is flat, which can affect the resolution of two close frequency components.

## 2.3.16  Calculation of the Signal Power

Often it is necessary to sum the power of the frequency components in a limited frequency range or in the entire spectrum. Using the time windows gives rise to sidebands, even in cases when they are not present in the spectrum which was weighted using the rectangular window.

The power of sinusoidal signals can also be calculated from the amplitude of this signal. If the amplitude is equal to $A$, then the root mean square ($RMS$) is equal to $A/\sqrt{2}$ and the power ($PWR$) is calculated as the square of $RMS$, therefore, $PWR = RMS^2 = A^2/2$.

Examples of the frequency spectrum which is calculated without weighing the input time signal and the spectra of the time signal which is weighted with the use of the time window

**Table 2.9**   Overview of the most used time windows.

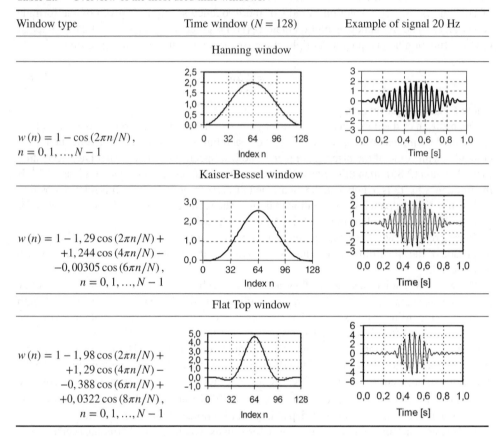

| Window type | Time window ($N = 128$) | Example of signal 20 Hz |
|---|---|---|

*Hanning window*

$w(n) = 1 - \cos(2\pi n/N)$,
$n = 0, 1, ..., N - 1$

*Kaiser-Bessel window*

$w(n) = 1 - 1,29\cos(2\pi n/N) +$
$+1,244\cos(4\pi n/N) -$
$-0,00305\cos(6\pi n/N)$,
$n = 0, 1, ..., N - 1$

*Flat Top window*

$w(n) = 1 - 1,98\cos(2\pi n/N) +$
$+1,29\cos(4\pi n/N) -$
$-0,388\cos(6\pi n/N) +$
$+0,0322\cos(8\pi n/N)$,
$n = 0, 1, ..., N - 1$

**Table 2.10**   Effect of the selection of the time window on the frequency spectrum.

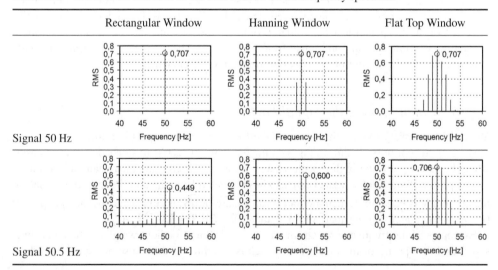

|  | Rectangular Window | Hanning Window | Flat Top Window |
|---|---|---|---|
| Signal 50 Hz |  |  |  |
| Signal 50.5 Hz |  |  |  |

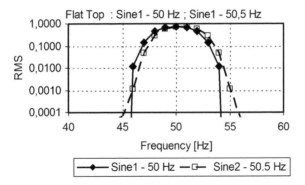

**Figure 2.29**   Effect of Flat Top time window on the frequency spectrum of the sinusoidal signals of 50 and 50.5 Hz.

of the rectangular or Hanning type are shown in Figure 2.30. In fact weighing with the use of the rectangular window is no weighting, because the time signal is not influenced. The power of the signal across the whole frequency range or only its parts can be calculated by integrating the power spectral density with respect to frequency. When using a time window, which changes the time signal, the result of the power calculation differs from the result you get when using the rectangular window or a simple calculation based on the amplitude of the sinusoidal signal.

This phenomenon is to be compensated for by using a parameter which is called the time window bandwidth and is denoted by *BW*. If we calculate the power *PWR* as the sum of the individual powers *PWR* (*k*) of the weighted power spectrum we get a higher power value than we obtain for the unweighted spectrum, that is, weighted formally only by a rectangular window. When using the power spectral density spectrum *PSD* (*k*) the introduction of the bandwidth *BW* describes the following equation for calculating the power of the signal

$$PWR \times BW = \sum_{k=0}^{k=N/2-1} PSD\,(k)\,\Delta f \Rightarrow PWR = \sum_{k=0}^{k=N/2-1} PSD\,(k)\,\Delta f/BW$$

$$(2.87)$$

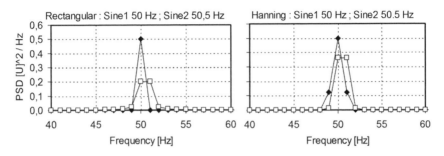

**Figure 2.30**   Power spectral density of the sinusoidal signals of 50 and 50.5 Hz.

**Table 2.11**   Calculation of the sum of the individual component powers.

| | Rectangular window | | | | Hanning window | | |
|---|---|---|---|---|---|---|---|
| Frequency | PWR | $\sum_k PSD(k)\Delta f$ | BW | Frequency | PWR | $\sum_k PSD(k)\Delta f$ | BW |
| 50 Hz | 0.5 | 0.500 | 1 | 50 Hz | 0.5 | 0.750 | 1.5 |
| 50.5 Hz | 0.5 | 0.499 | 1 | 50.5 Hz | 0.5 | 0.750 | 1.5 |

An example with two sinusoidal signals of the frequency 50 and 50.5 Hz is shown in Figure 2.29 and the calculation of the power in Table 2.11. Bandwidth for the other time windows is given in Table 2.12.

## 2.3.17   Frequency Weighting

The human ear is more sensitive to frequencies between 500 Hz and 8 kHz than the very low and high frequencies. Measuring microphones as a sensor of sound pressure are compared to human ears as being equally sensitive in the frequency range from 20 Hz to 20 kHz. To ensure that the measuring device indicates pretty much what you actually hear we have to use a frequency filter to emphasise or suppress some partial frequency ranges. The microphone is a sensor of the sound pressure in Pa. In practice we use dimensionless decibels (dB) which are defined as 20 times the logarithm of the ratio of the sound pressure in RMS relative to the reference level of 2 x $10^{-5}$ Pa called the threshold of hearing at 1 kHz. For sound the unit of loudness is the phone which is a numerical equivalent to the sound pressure level in dB only at 1 kHz. There are at least three types of filters, namely for A-weighting, B-weighting and C-weighting.

The A-weighting filters out significantly more bass than the others, and is designed to approximate the ear at around the 40 phon level. It is very useful for eliminating inaudible low frequencies. The intermediate B-weighting approximates the ear for medium loud sounds. It is rarely used. The C-weighting does not filter out as much of the lows and highs as the other contours. It approximates the ear at very high sound levels and has been used for traffic noise surveys in noisy areas. The frequency responses of the above mentioned filters are shown in Figure 2.31.

**Table 2.12**   Time window bandwidth.

| Time window | Bandwidth BW |
|---|---|
| Rectangular | 1 |
| Hanning | 1.5 |
| Kaiser-Bessel | 1.8 |
| Flat Top | 3.77 |

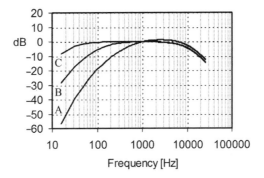

**Figure 2.31**   Frequency weighting functions A, B and C.

The weighting function of the A type is the most widely used in practice. The A-weighted spectrum $S_A(f)$ in RMS of the sound pressure is given by

$$S_A(f) = H_A(f) S(f) \tag{2.88}$$

where $S(f)$ is the unweighted RMS spectrum of the sound pressure and $H_A(f)$ is the magnitude of the frequency response of the A type filter. To simplify the following formulas we will define $f_1 = f/1000$. The magnitude of the frequency response is as follows

$$H_A(f) = \frac{K\left(f_1^2\right)^2}{\left(\left(f_1^2 + a_1\right)f_1^2 + a_2\right)\sqrt{\left(f_1^2 + a_3\right)f_1^2 + a_4}} \tag{2.89}$$

where   $K = 187.3742$,   $a_1 = 148.8404$,   $a_2 = 0.06316174$,   $a_3 = 0.5560957$,   and   $a_4 = 0.006315772$.

Weighting function suppresses low-frequency components of the spectrum. An example of howthe low frequency components of the noise produced by a gearbox are affected is shown in Figure 2.32. The low frequency components are excited in the case of gearboxes by the shaft rotational frequency and its harmonics.

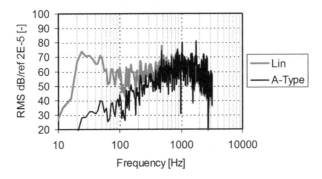

**Figure 2.32**   Weighted (A-Type) and unweighted (Lin) RMS spectrum of noise produced by a gearbox.

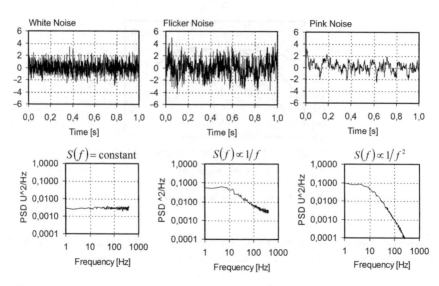

**Figure 2.33**   Time history and PSD spectrum of white noise, flicker noise and pink noise.

### 2.3.18   Analysis of Random Signals with the Use of Averaging in the Frequency Domain

In practice we measure random signals in additionto deterministic signals.. Deterministic signals are composed of several sinusoidal signals and their spectrum therefore conveniently describes the scale in RMS. In contrast to this, the random signals best describe the power spectral density. If the signal units are generally designated by (U) then the units of the PSD spectrum are (U²/Hz). Purely random signals do not contain dominant frequency components in their frequency spectrum. The previously mentioned power spectral density of these signals depends on the frequency. Power spectral density of the white noise does not depend on frequency. They are also random signals of the $1/f^{\alpha}$ type. For pink noise the parameter $\alpha$ equals to 2 while for flicker noise $0 < \alpha < 2$ and for white noise $\alpha = 0$.

Examples of the time history and the corresponding PWR spectrum of the various types of random processes are shown in Figure 2.33. The spectra correspond to the 50 second record of the signal, while the time signals show only one second is needed to demonstrate their appearance in detail. The method of calculating a representative spectrum will be described in this section.

Statistics of random variables deal with statistical samples arranged into a random vector which is selected from a theoretically infinite number of possible values of a random variable. We will focus our interests on the mean value of the random variable. For the finite number of samples only estimates of the mentioned statistical parameter which is close to the unknown exact mean value can be calculated. The estimate of the mean value is in fact a random variable. The standard deviation of the mean value as a random value is proportional to the reciprocal of the square root of the number of the samples. Increase in the number of samples leads to a more accurate estimation of their mean value.

The mean of the set of the statistical samples in the sense of maximum likelihood is given by the arithmetic average of the individual samples or in other words, by the sum of the weighted samples. Each value of $X_k$ in the calculation of the DFT using the formula (2.55) is also a weighted average, however, also a complex number because the weights are complex. Neither the values of $X_k$ nor the RMS of the corresponding spectrum component are suitable for direct averaging. The signal power at the $k$-th frequency is given by $X_k X_k^* = |X_k|^2$, which is a positive real number. In the same way as the mean value of samples is estimated an average power can be estimated at the $k$-th frequency from the finite sequence of samples.

Averaging will be described for the power spectral density. Each of the instantaneous spectra is designated by $psd_m, m = 1, 2, ..., M$ where $M$ is the total number of the instantaneous spectra. The integer number $M$ is known as the number of averages. Each of the $k$-th components of the averaged spectrum is calculated using the following formula

$$PSD_M(k) = \frac{1}{M} \sum_{m=1}^{m=M} psd_m(k), \quad k = 0, 1, 2, ..., N/2 - 1 \qquad (2.90)$$

where $N$ is the length of the record for calculating DFT. The calculation procedure is shown in Figure 2.34.

Since the input signal to calculate the spectrum in Figure 2.34 is of the white noise type then the spectrum resulting from an averaging process is almost a constant, that is, the spectrum is

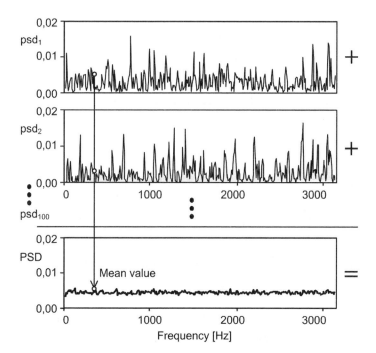

**Figure 2.34**   Calculation of the averaged spectrum of the white noise signal.

independent on the frequency. Averaging the instantaneous spectra is a way of smoothing the spectrum of the random signals.

Repeated calculations of the mean values by the formula 2.78 can be replaced by a recursive procedure for the calculation of $PSD_M$ which benefits from the previous results of $PSD_{M-1}$ calculation

$$PSD_M(k) = \frac{M-1}{M} PSD_{M-1}(k) + \frac{1}{M} psd_M(k), \quad k = 0, 1, 2, \ldots, N/2 - 1. \quad (2.91)$$

The signal of a noise or vibration may be a mixture of sinusoidal signals and random measurement errors in the meaning of the background of the RMS spectrum. In this case we prefer the RMS scale to the PSD scale. Averaging the RMS spectra is started up by the conversion of the RMS scale to the PWR scale, for which is the averaged PSD spectrum is calculated and then this PWR spectrum is re-converted back to the RMS spectrum.

$$rms_m \rightarrow psd_m = rms_m^2/\Delta f \rightarrow \text{averaging } psd_m \rightarrow PSD \rightarrow RMS = \sqrt{PSD\Delta f}. \quad (2.92)$$

So far, we have only described the method for averaging the power spectral density without selection of the signal segments of the length $N$ for calculation of DFT.

There are many ways of associating the segments of the unified length with the time signal as is shown in Figure 2.35. If the adjacent segments do not overlap, then there is a loss of information about the signal. It seems that the direct connection of the segments with the overlap of 0% is without loss of information. But the time window may reduce the weight signal to zero at the connection of the adjacent segments. Therefore it is used as an overlap of 50 or 66%, that is, the segments overlap by half or by two thirds of their length.

The segments are weighted by the square of the weighting function of the Hanning type. The total weighting function of the overlapping segments is illustrated in Figure 2.36. Except for the overlap of 66% the other weighting functions are not a constant and are independent of time. It can be shown that only for the overlap $n/(n + 1) \times 100\%$, where $n$ is an integer greater than unity, is the weighting function a constant.

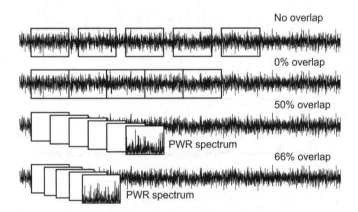

**Figure 2.35**   Overlap of records for calculation of DFT.

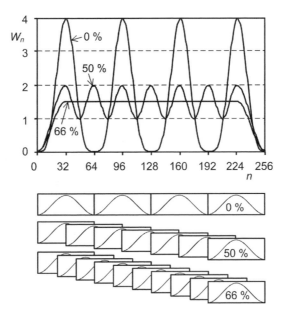

**Figure 2.36** Effect of the Hanning time window on the averaged spectrum.

The effect of the number of averages on the relative estimation error of the averaged power spectral density determines the following formula

$$\varepsilon_r = \frac{1}{2\sqrt{M}}. \tag{2.93}$$

An example which shows the dependence of the averaged spectrum of white noise on the number of averages is illustrated in Figure 2.37. The table in this figure shows the relationship between the number of averages and the length of the time interval for calculation of the averaged PSD spectrum. The frequency spectrum of the white noise is theoretically indepen-dent of the frequency and is equal to a constant value, however, for a time interval which is theoretically infinitely long. The larger the time interval, the smaller the standard deviation of the spectrum values from the mean. The above described method of calculation of spectra is also known as Welch's method.

The signal analysers which are designed for real-time operation are displaying continuously the running results of averaging. This averaging has three modes:

- exponential;
- linear;
- peak.

The recursive formula (2.91) is used for averaging in the exponential and linear mode. The number of averages is a part of the settings of the analyser for measurement conditions. The actual number of averages in the formula (2.91) over the process of averaging is changed from

| Averages | Time interval [s] |
|----------|-------------------|
| 1        | 0.128             |
| 2        | 0.171             |
| 5        | 0.299             |
| 10       | 0.512             |
| 20       | 0.939             |
| 50       | 2.219             |
| 100      | 4.325             |

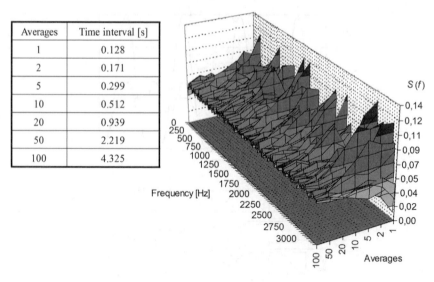

**Figure 2.37**   Effect of the number of averages on the averaged spectrum.

one to the preset value. The number of averages for the first calculation of the spectrum is set to unity, and then gradually increases with an increment of one to the preset value. The averaged spectrum in the first step of averaging is not calculated, but taken directly from the calculation of the instantaneous spectrum.

Averaging and measurements in the linear mode is stopped when the actual number of averages reaches the preset value. In contrast to the linear mode the averaging and measurements in the exponential mode continues after reaching the preset value of the number of averages and will remain a constant value for calculating the formula (2.91).

The averaged spectrum in the peak mode is not calculated according to the formula (2.91) but presents the largest amplitude of each spectral line. After calculating a new instantaneous spectrum the magnitude of the spectrum components is compared with the stored magnitudes. If the new magnitude is greater then the previously stored one then this magnitude is replaced by the greater magnitude. Measurement stops on reaching the number of averages. This method creates the spectrum envelope.

## 2.4   Zoom FFT

The frequency analysis in the baseband frequency range which begins with a frequency of zero needs to be described. It is sometimes important to know the spectrum with high resolution in a band around the centre frequency which is distant from zero frequency. The reason is that the sideband components of the centre frequency component contain a lot of useful information. The frequency resolution of the FFT spectrum in the baseband range is given by the formula $\Delta f = 1/T$, where $T = N/f_S$. The length of the input sequence for calculation of FFT is equal to the integer number $N$. The sampling frequency is as usual designated by $f_S = 1/T_S$. If it the finer resolution over a limited portion of the spectrum is required then it is possible in a baseband range to apply the so-called Zoom-FFT. Zooming-in on the aforementioned portion of the frequency spectrum is defined by the central frequency of $f_C$ and the bandwidth of $\Delta F$, which contains the same number of lines as the normal baseband spectrum.

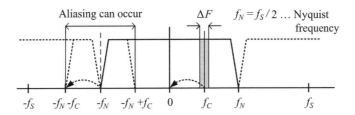

**Figure 2.38** Frequency shift due to the multiplication by a rotating unit complex vector.

Increasing the resolution of the frequency spectrum requires the time record of the sampled signal for FFT calculation to be extended and at the same time it is necessary to prepare data so that the calculation result concerns only a narrowband frequency range. Two calculation procedures of the zoom FFT were described by Randall in his book on frequency analysis. These methods are named as follows:

- real-time zoom;
- nondestructive zoom.

## 2.4.1   Real-Time Zoom

The method, called real-time zoom, is based on the shift of the frequency origin to the centre of the zoom band. Multiplication can only be done for the sampled signal. The real samples $x_n, n = 0, 1, 2, \ldots$ are multiplied by a rotating unit complex vector $\exp\left(-j2\pi f_C nT_S\right)$

$$y_n = x_n \exp\left(-j2\pi f_C nT_S\right), \quad n = 0, 1, 2, \ldots \tag{2.94}$$

The sequence of the real samples $x_n, n = 0, 1, 2, \ldots$ is transferred into the sequence of the complex samples $y_n, n = 0, 1, 2, \ldots$. Multiplication by the rotating vector may cause aliasing in the negative frequency band as is shown in Figure 2.38. The two-sided frequency spectrum is used to demonstrate the frequency shift. After the frequency shift the portion of the frequency spectrum with negative frequencies ranging from $-f_N$ to $-f_N + f_C$ will have a larger absolute value than the Nyquist frequency of $f_N$, and therefore there is a risk of mirroring the component of the frequency which ranges $-f_N - f_C$ to $-f_N$ around the frequency of $-f_N$. A low-pass filtering with a cut-off frequency of $\Delta F/2$ has to be used to prevent aliasing.

A digital FIR (Finite Impulse Response) filter in the time domain can be used, as it has a linear phase as shown in Figure 2.39. This type of filter can be designed with a very short transient frequency band. The real and imaginary samples are filtered separately. The sampling frequency should be tailored for the frequency range. The original sampling frequency may be too high for the signal at the output of the low pass filter and therefore this sampling rate can

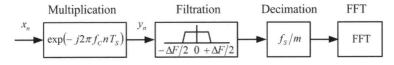

**Figure 2.39** Real-time zoom method.

be reduced $m$ times. Reducing the sampling frequency is possible using the decimation that is, regular omission of certain samples. Omitting every other sample will reduce the frequency of sampling by two. Reducing the sampling frequency $m$ times means the selection of each $m$-th sample. The parameter $m$ is called a zoom factor. The stream of the output complex samples after decimation can be processed using the FFT algorithm. The frequency range from $-\Delta F/2$ to $+\Delta F/2$ in the displayed spectrum is replaced by the original frequency range of the $\Delta F$ bandwidth around the centre frequency of $f_C$, that is, from $f_C - \Delta F/2$ to $f_C + \Delta F/2$.

### 2.4.2 Non-Destructive Zoom

According to the nondestructive zoom method it is necessary to process the extended record of the length of $mN$, where $m$ is the zoom factor and $N$ is a record length, for calculation of FFT. The high-resolution frequency spectrum is obtained thanks to an extra long sequence of input samples to calculate the FFT. The following description shows how to divide the calculation of FFT into many calculations for much shorter sequences.

The extended record of length $mN$ is divided into the $m$ sub-sequences of length $N$ as follows

$$
\begin{aligned}
x_{0+\mu m}, & \quad \mu = 0,\ 1,\ 2, \ldots,\ N-1, \\
x_{1+\mu m}, & \quad \mu = 0,\ 1,\ 2, \ldots,\ N-1, \\
\ldots & \quad \ldots \\
x_{m-1+\mu m}, & \quad \mu = 0,\ 1,\ 2, \ldots,\ N-1.
\end{aligned}
\tag{2.95}
$$

The individual sub-sequences are delayed by $n\,T_S$, $n = 0,\ 1, \ldots,\ m-1$ compared with the first sub-sequence. FFT is possible to calculate using the sequence of the length of $mN$. However, this calculation can be replaced by the repeated calculations of FFT for the input sub-sequence of the length of $N$. In this case it is necessary to repeat the calculation of m times. The calculation of the sum can be decomposed into the sum of the particular sums according to the diagram in Figure 2.40

$$
X_k = \sum_{n=0}^{mN-1} x_n \exp\left(-j\frac{2\pi}{mN}kn\right) = \sum_{v=0}^{m-1}\sum_{\mu=0}^{N-1} x_{v+\mu m} \exp\left(-j\frac{2\pi}{mN}k\,(v+\mu m)\right)
$$

$$
= \sum_{v=0}^{m-1} \exp\left(-j\frac{2\pi}{mN}k\,v\right) \left(\sum_{\mu=0}^{N-1} x_{v+\mu m} \exp\left(-j\frac{2\pi}{N}k\,\mu\right)\right)
\tag{2.96}
$$

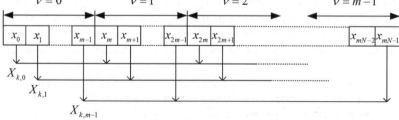

**Figure 2.40** Non-destructive zoom method.

It is possible to define a term for each sub-sequence which is indexed by $k$ and $v$ as follows

$$X_{k,v} = \left( \sum_{\mu=0}^{N-1} x_{v+\mu m} \exp\left(-j\frac{2\pi}{N}k\,\mu\right) \right) \tag{2.97}$$

It is clear that $X_{k,v} = X_{k+N,v}$. Therefore the term $X_{k,v}$ is $N$-periodic with respect to the index $k$. In this case it is sufficient to calculate the value of $X_{k,v}$ only for $k = 0, 1, 2, ..., N-1$ and not for $k = N, N+1, ..., mN-1$.

The partial results are put together by the formula

$$X_k = \sum_{v=1}^{m} \exp\left(-j\frac{2\pi}{mN}kv\right) X_{k,v}, \quad k = 0, 1, 2, ..., mN-1. \tag{2.98}$$

The values of $X_k$ can be computed for all possible values of the index $k$ or only for certain values that belong to the selected frequency band.

## 2.5 Filtration in the Frequency Domain

There are two methods of modifying the frequency spectrum of a time signal. The first method is based on the use of the analogue or digital filters that process the signal in the time domain and the second method uses the frequency domain to modify the spectrum of signals [3]. The analogue filters can process the time signals only in real-time, all the analogue filters are causal. The digital filter is either causal or not-causal. The causal digital filter processes only the previously measured samples while the not-causal filter can use all the samples as an input, either from the past or future only if they are stored in memory for post processing. The issue of digital filtering, however, is not addressed in this book.

The modification of the signals in the frequency domain is used in the signal analysers which are based on FFT, and hence the input data for calculating the frequency spectrum is a sequence of the samples of the finite length. These analysers operate in real-time and are designed to measure vibration spectra in the form of velocity or acceleration and noise spectra. If there a signal of acceleration is measured and the frequency spectrum of the velocity signal is required, then it is necessary to integrate acceleration with respect to time for calculating the velocity spectrum and vice versa for calculating the acceleration spectrum. The functionality of integration and differentiation is part of the signal analyser software. In addition to integration and differentiation with respect to time the frequency domain may be used for removing interfering components or for analysing the components in the frequency range of interest. The records of measurements on the transmission units of vehicles can be processed in intervals corresponding to one complete revolution of the selected components. Filtering in the frequency domain for this is very useful.

Filtering in the frequency domain relates only a finite sequence of the samples of the real-valued signal $x_n, n = 0, 1, ..., N-1$. These samples are transformed into the frequency domain using the discrete Fourier transform. The sequence of complex numbers $X_k, k = 0, 1, ..., N - 1$ of the same length as the input sequence of the real numbers is the result of this transformation.

$$\mathbf{X} = DFT\{\mathbf{x}\}. \tag{2.99}$$

where $\mathbf{x} = [x_0, x_1, ..., x_{N-1}]^T$ and $\mathbf{X} = [X_0, X_1, ..., X_{N-1}]^T$ are vectors. The complex numbers $X_k$ can be considered as a component of the Fourier spectrum. The components are indexed so that it starts at zero and ends at $N - 1$. The component frequency is an integer multiple of the fundamental frequency $f_1 = 1/T$ where $T$ is a period (s) that corresponds to the length of the input sequence.

The pair of the complex values $X_k$ and $X_{N-k}$ for $0 \le k \le N/2$ determines the amplitude and phase of the sinusoidal signal components of the frequency $f_k = k/T$ where $0 \le k \le N/2$. The second frequency $f_k = (N - k)/T$ of the mentioned pair is formally greater than the Nyquist frequency, therefore it is more suitable as a designation of the negative frequency. If the index $k$ is less or equal to half the number $N$ of samples then it is preferable to use the designation negative frequency $f_{-k} = -k/T$ which has an absolute value less than the Nyquist frequency. As explained earlier the components $X_k$ and $X_{N-k}$ are the complex conjugate, and therefore have the same absolute value and opposite phase. This is the reason for displaying only one-sided spectra.

Filtering is a modification of the components $X_k, k = 0, 1, ..., N - 1$ to other components $Y_k, k = 0, 1, ..., N - 1$ so that the output from the inverse Fourier transform of these modified components is again a real signal. In order to preserve this property, the components which are indexed by $k$ have to be a complex conjugate value of the components which are indexed by index $N - k$. This rule does not apply to indices 0 and $N/2$. The change of the values of these complex numbers can result in the filtration, integration and calculation of derivatives. After modifying, the sequence of the complex numbers $\mathbf{Y} = [Y_0, Y_1, ..., Y_{N-1}]^T$ is transformed back into the time domain by using the inverse Fourier transform.

$$\mathbf{y} = IDFT\{\mathbf{Y}\}. \tag{2.100}$$

### 2.5.1   Filtering

Filtering is a process that removes some unwanted component or a group of components from a signal. This removal is achieved if this pair of components is set to zero

$$Y_k = Y_{N-k} = 0, \quad 0 < k < N/2 \tag{2.101}$$

while others are assigned to the modified Fourier transform without changing $Y_k = X_k$. The DC component of the signal is removed by setting the component $Y_0$ to zero. By setting some components to zero any kind of filter can be realised, for example, a low pass, high pass, band pass and band stop filter. This technique allows a comb filter to be created. It is possible to substitute zero for only one component and in this way to remove it from the signal. If this component is somehow modulated, substitution zero is also necessary for several adjoined sideband components.

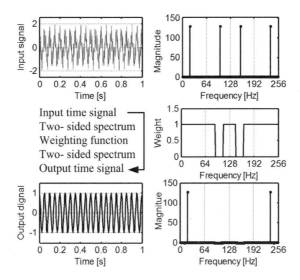

**Figure 2.41**  Band stop filtration in the frequency domain.

An example of filtering in the frequency domain is shown in Figure 2.41. The component of the frequency of 100 Hz is removed from the signal. The advantage of this filtering process is that the output signal is free of phase delay which is usually found when using the digital filters. To emphasise the need to modify the two components which are symmetric around the Nyquist frequency the Fourier spectrum is displayed as the two-sided spectrum.

## 2.5.2   Integration and Calculation of Derivatives

Integration and differentiation with respect to time in the time domain is replaced by the elementary mathematical operations, such as division or multiplication, in the frequency domain. The original Fourier transform is divided by the term $j\omega$ if the signal is integrated in the frequency domain. This original Fourier transform is multiplied by the term $j\omega$ if it is calculated as the first derivative of the signal in the frequency domain. With regard to the complex conjugation of the components pairs, the frequencies lower and higher than the Nyquist frequency need to be distinguished for calculations in the frequency domain.

The general formula for integration in the frequency domain has to respect the complex conjugate symmetry of the result. The algorithm for calculating the individual components is as follows

$$
\begin{aligned}
&Y_0 = Y_{N/2} = 0 \\
&k = 1, 2, \ldots, N/2 - 1: \\
&\quad Y_k = X_k/(j2\pi k f_1) \\
&\quad Y_{N-k} = -X_{N-k}/(j2\pi k f_1).
\end{aligned}
\tag{2.102}
$$

where $f_s = N/T$ is the sampling frequency.

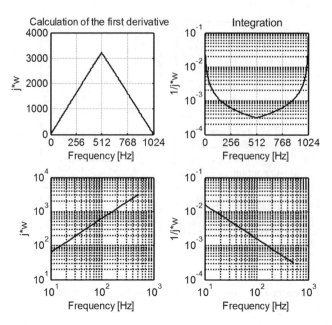

**Figure 2.42**   Absolute value of the weighting function for calculation of derivatives and integration.

The general formula for calculation of the first derivative with respect to time in the frequency domain has to fulfil the same condition. The algorithm for calculating the individual components is as follows

$$
\begin{aligned}
Y_0 &= Y_{N/2} = 0 \\
k &= 1, 2, \ldots, N/2 - 1 : \\
Y_k &= j2\pi k f_1 X_k \\
Y_{N-k} &= -j2\pi (N - k) f_1 X_{N-k}.
\end{aligned}
\tag{2.103}
$$

Expression $j\omega$ is the weighting factor which depends on the frequency. This functional dependence can be represented graphically as in Figure 2.42. The absolute value of this factor is shown in the diagram for simplicity.

An example for integration and calculation of the first derivative of a signal is shown in Figure 2.43. The input signal is composed of two harmonics. Both mathematical operations affect the amplitude of the harmonics in the output signal. Integration acts as a low pass filter. The DC component of the output signal is zero. The calculation of the first derivative corresponds to the filter whose gain is proportional to the frequency. The phase of each component in the output signal after the calculation of the first derivatives is shifted by the $\pi/2$ radians. The measurement errors as a high frequency noise are a part of each measured signal, and therefore the result of calculation can be spoiled by immersing the useful signal in random noise. Therefore it is essential to combine the calculation of the derivative with the filtering of high frequency components. The integration of the acceleration signal is strongly influenced by the low frequency drift of acceleration sensors. In this case, it is appropriate to remove these low frequency components in the input signal.

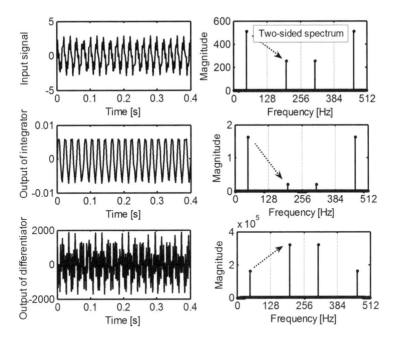

**Figure 2.43**  Weighting function for calculation of derivatives and integration.

When calculating the double integration and double derivatives the term $j\omega$ is replaced by the expression of $(j\omega)^2 = -\omega^2$.

## 2.6  Average Power of the Signal

The average power of the signal over a time interval which is needed to record a certain number of samples to calculate the Fourier transform can be calculated by the formula (2.90). The root mean square (*RMS*) of a signal is a root of this signal power. The root mean square (*RMS*) of the signal can be calculated as the root of the signal power (*PWR*). The calculation result which is based on the frequency spectrum is known in some signal analysers as a parameter called TOTAL in either *RMS* or *PWR* depending on what type of the scale is displayed. The instantaneous power of the time signal is given by the square of the signal samples $x_n^2$, $n = 0, 1, 2, \ldots$. The average power of the signal over the $n$ signal samples is given by the formula

$$PWR_n = \frac{1}{n} \sum_{i=1}^{n} x_i^2. \tag{2.104}$$

If the new sample $x_{n+1}$ is measured, the new power is adjusted to this sample by the formula

$$PWR_{n+1} = \frac{1}{n+1} \sum_{i=1}^{n+1} x_i^2 = \frac{1}{n+1} \left( \sum_{i=1}^{n} x_i^2 + x_{n+1}^2 \right) = \frac{1}{n+1} \left( n \times PWR_n + x_{n+1}^2 \right)$$

$$= \frac{n}{n+1} PWR_n + \frac{1}{n+1} x_{n+1}^2 \tag{2.105}$$

Because the expression of $n/(n+1)$ can be rewritten in the form of $1 - 1/(n+1)$ the new power of the signal is calculated as a weighted average of the previous power $PWR_n$ and the new instantaneous power $x_{n+1}^2/(n+1)$. The sum of the two weights is equal to one. The fraction of $1/(n+1)$ is the same as the ratio of the sampling period $T_S$ to the chosen length of the time interval, designated as a time constant $\tau$, over which the power of the signal is calculated. The formula for calculating the average power is as follows

$$PWR_k = \left(1 - T_S/\tau\right) PWR_{k-1} + x_k^2 T_S/\tau. \tag{2.106}$$

The time constant has two specially named sizes, SLOW for $\tau = 1\,\text{s}$ s and FAST for $\tau = 1/8$s. There is no problem in presenting the RMS dependence on time in a similar way to the signal power dependence on time. The function of calculating the signal power is referred to as an overall instrument in some signal analysers.

# References

[1] Cooley, J.W. and Tukey, J.W. (1965) An algorithm for machine calculation of complex Fourier series. *Math of Computation*, **19**, 297–301.
[2] Oppenheim, A.V., Schafer, R.W. and Buck, J.R. (1999) *Discrete-Time Signal Processing*, 3rd edn, Prentice Hall Inc., Upper Saddle River, New Jersey.
[3] Mitra, S.K. (2006) *Digital Signal Processing: A Computer Based Approach*, 3rd edn, McGraw Hill Publications, New York.
[4] Randall, R.B. (1987) *Frequency Analysis*, Brüel & Kjaer, revision September 1987.
[5] Suranek, P., Mahdal, M., and Tuma, J. (2013) *Comparising antialiasing filters in A/D converters*. 14th International Carpathian Control Conference ICCC 2013, May 26–28, 2013, Rytro, Poland, pp. 373–376.

# 3

# Gearbox Frequency Spectrum

## 3.1 Source of Gearbox Noise and Vibration

The main source of noise and vibration are the gears and bearings. Plenty has been written about the vibrations which are excited by bearings [1]. This book is primarily concerned with gears of the involute type which transfer the angular velocity of the driving gear to the angular velocity of the driven gear exactly according to the gear ratio and this gear ratio is insensitive to the distant of the centres of gears in mesh. Due to the final compliance of teeth in mesh the stiffness of the teeth is more complicated that it seems to be. Conversion of the angular velocity is not as simple for the loaded gear train as for the unloaded gear and as a result a transmission error arises.

Firstly, we describe how vibrations arise. A gear train can be in several operating modes. One of these modes is the condition where the gear train is not transmitting the torque and the teeth move within their backlash range. This operation mode produces a gear rattle. This is not the main operating mode of the gearbox. For a description of the main sources of noise it will be assumed that the teeth are in mesh. Each pair of teeth first makes contact at a single point and then they slide over each other in the axial direction. The entrance into the mesh may cause a small mechanical impact which excites structural vibration as it encounters defects on the raceway of the rolling bearings. These vibrations are transmitted by the gearbox structure. We can measure the frequency of these pulses, but the structural resonance is not usually identified. There are many resonance frequencies of the gearbox structure, but they are not associated with an individual part of the gearbox or a gear. The mechanical shock is not accompanied by damped oscillations as in the case of defects in rolling bearings.

Vibrations of the gearboxes can sufficiently explain a phenomenon which is called a parametric excitation. If the individual teeth alternate in mesh then the stiffness of toothmeshing changes according to the number of tooth pairs in mesh. Moreover the point where are the tooth flanks touch moves in the radial direction. Without a doubt, a step change of stiffness occurs when the number of the pairs of teeth in meshes changes. For spur gear one or two pairs of teeth are alternately in mesh. An example which demonstrates the dependence of stiffness on the angle of rotation is in Figure 3.1. In this figure the dimensionless parameter $\varepsilon_\alpha$ which is

*Vehicle Gearbox Noise and Vibration: Measurement, Signal Analysis, Signal Processing and Noise Reduction Measures*, First Edition. Jiří Tůma.

Stiffness of a single tooth pair

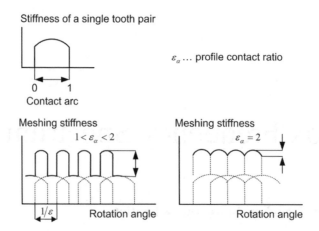

$\varepsilon_a$ ... profile contact ratio

**Figure 3.1**   Toothmeshing stiffness.

called a profile contact ratio is introduced. This approximately indicates the average number of teeth in mesh during meshing cycles.

The dependence of the toothmeshing stiffness on the angle of rotation for two values of the profile contact ratio is shown in Figure 3.1. The teeth of the gears with the profile contact ratio between 1 and 2 is called a low contact ratio (LCR) and teeth of the gears with the profile contact ratio equal to 2 is designated as a high contact ratio (HCR). Without calculation it is obvious that the gear meshing stiffness variations are smaller and therefore cause less parametric excitation of gearbox vibrations.

There are various academic mathematical models describing dynamic behaviour of the gear in mesh. One of the simplest is designed for a kinematic scheme in Figure 3.2. The equation

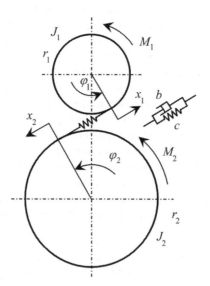

**Figure 3.2**   Kinematic scheme of the gear train.

of rotational motion of two gears which are connected by a spring-damper system is as follows

$$J_1 \ddot{\varphi}_1 + b(t)\left(\dot{\varphi}_1 - \dot{\varphi}_2\right) + c(t)\left(\varphi_1 - \varphi_2\right) = M_1$$
$$J_2 \ddot{\varphi}_2 + b(t)\left(\dot{\varphi}_2 - \dot{\varphi}_1\right) + c(t)\left(\varphi_2 - \varphi_1\right) = M_2.$$

(3.1)

where $M_1$ and $M_2$ are torques (moments of force), $\varphi_1$ and $\varphi_2$ are angles of rotation, $J_1$ and $J_2$ are moment of inertia, $b$ and $c$ are damping and stiffness of the tooth contact, respectively. Equations of motion can be modified into the reduced form for modelling the deformation of the teeth

$$x = x_1 + x_2 = r_1 \varphi_1 + r_2 \varphi_2$$

(3.2)

where $r_1$ and $r_2$ are radii of the pitch circle. The reduced form of the equation of motion is as follows

$$m_{RED} \ddot{x} + b(t)\dot{x} + c(t)x = M_1/r_1 - M_2/r_2$$

(3.3)

where $m_{RED}$ is a reduced mass

$$m_{red} = \frac{J_1 J_2}{J_1 r_2^2 + J_2 r_1^2}.$$

(3.4)

For equilibrium state between the input and output torque

$$M_2 + M_1 \frac{r_2}{r_1} = 0$$

(3.5)

the right side of the reduced equation of motion becomes zero. Because the right side of the differential equation is equal to zero the vibrations are not forced or free, but self excited. Coefficients of damping and stiffness are periodic functions of time that excite periodic solution of the nonstationary differential equation. The coefficients and the equation solutions have a period $T_{GMF} = 1/f_{GMF}$ where $f_{GMF}$ is the gear mesh or toothmeshing frequency. The periodical deformation of the spring causes angular vibrations of the shafts with the mounted gears. The angular vibrations and inertia of the gears cause the dynamic force acting between the teeth [2].

Mechanical power is transmitted from a driving gear to the driven gear by means of the force $F_T$ acting along the line of action. This force is compensated by the same and antiparallel force $F_S$ acting at the shaft support point. The simultaneous acting of these forces $F_T$ and $F_S$ to the driven gear results in torque, see Figure 3.3. The teeth contact stiffness is not equal to a constant value, but it is oscillating in synchronism with toothmeshing frequency due to the oscillation of the number of the tooth pairs in contact and moving the teeth contact point along the tooth flank. The oscillation of the tooth contact stiffness causes the self excited angular vibration of the driven gear, which results in the time varying forces $F_T$ and $F_S$. The force acting at the shaft support is dynamic as well and excites vibration of the gearbox housing and consequently emitting noise.

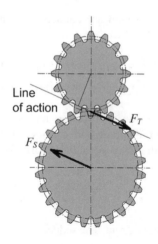

**Figure 3.3**   Transmitting angular vibration to the gearbox housing vibrations.

The contact of engaged teeth is not a point but a straight line. The distribution of forces along this line may be non-uniform. Analysis of the solution of the reduced differential equation is an academic issue, the test engineer or designer prefers an experiment. There could be an entire book devoted to experiments with measuring vibrations of the gearbox housing, noise radiated by the gearbox and angular vibrations of the gearbox shafts. The relationship between the angular vibrations and the linear vibrations of the gear train will be illustrated by the example in Figure 3.4. Both signals which are acceleration have the same frequency range up to one hundred times the speed of rotation. The linear acceleration is measured by an accelerometer which is placed on the bearing of the shaft where is the gear is mounted and is orientated in a perpendicular direction to the shaft axis. Angular acceleration is measured by an incrementary rotary encoder (IRC). A description of the method of calculating the angular acceleration and how to assign records to the selected gear are included in this book.

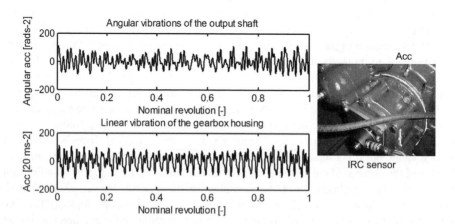

**Figure 3.4**   Transmitting angular vibration to the gearbox housing vibration.

## 3.2  Spectrum Signature

By definition, gears are provided with teeth on their perimeter, which serve to transmit torque to another gear or a rack. Teeth ensure rotation without slip. Firstly, it is supposed that gear axes do not move. As already mentioned, there are many types of gearings listed according to the relative position of the gear axes and the shape of the teeth, but for our purposes this is irrelevant. We will consider only frequency of repeating mesh cycles.

Gearboxes are machines running in cyclic fashion which results in the fact that noise produced by a gearbox is tonal. This means that the noise frequency spectrum consists of sinusoidal components at discrete frequencies with low-level random background noise. Because the emitted noise originates in the gearbox vibration, the vibration frequency spectrum has the same composition as the noise spectrum. The basic frequency of the rotating machines is the rotational speed $f_0$ given by the number of revolutions per second, in other words, in Hz. Another measurement for machine speed is the number of revolutions per minute, which is denoted by the abbreviation RPM (Revolutions Per Minute). The frequency of teeth impact or oscillation of the tooth-contact stiffness is computed as a product of the gear rotational speed in Hz and the number of teeth $n$ are referred to as the fundamental toothmeshing frequency or gear meshing frequency $f_{GMF}$

$$f_{GMF} = nf_0. \tag{3.6}$$

A gear pair in mesh forms a simple gear train which is characterised by only one toothmeshing frequency. The pair of gears in mesh may be extended optionally by an idler gear, which is an intermediate gear inserted between both the gears while meshing with them separately. The intermediate gear does not affect the toothmeshing frequency.

Compound gear trains consist of several pairs of meshing gears as can be seen in Figure 3.5. The compound gear train consists of two parallel shafts, the pair of gears in mesh and the gear, which is in mesh with the gear which is not shown in Figure 3.5. The first gear of the number of teeth $n_1$ is mounted on the primary shaft while the second gear of the meshing pair of the number of teeth $n_2$ is mounted on the secondary shaft. Both the gears in mesh are external what means that the teeth are formed on the outer surface of a cylinder or cone. The third gear of the number of teeth $n_3$ is mounted on the second shaft as well. The rotational speed of the first shaft is denoted by $f_1$ and the rotational speed of the second shaft is denoted by $f_2$.

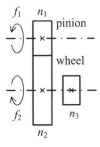

**Figure 3.5**   Compound gear train.

Considering the direction of gear rotation the rotational speed of the second shaft is given by the formula as follows

$$f_2 = -f_1 n_1 / n_2. \tag{3.7}$$

The minus sign in the formula means that the second gear rotates in the opposite direction. The ratio of the tooth number $n_1$ to the tooth number $n_2$ is called a gear ratio of the gear train. The gear of the gear train that has fewer teeth than the meshing gear is called pinion while the second gear of the gear train with more teeth than the pinion is called a wheel.

The toothmeshing frequency of all the gears of the gear train in Figure 3.5 is given by the formula as follows

$$
\begin{aligned}
f_{n1} &= n_1 f_1, \\
f_{n2} &= n_2 f_2 = -n_1 f_1 = -f_{n1}, \\
f_{n3} &= n_3 f_2.
\end{aligned}
\tag{3.8}
$$

For an arbitrary number of teeth $n_1$ and $n_2$ the absolute value of the toothmeshing frequency of both the gears in mesh is the same, while for different numbers of teeth $n_2$ and $n_3$ the toothmeshing frequencies of the corresponding gears are different. A sign of the meshing frequency is not important for one-sided frequency spectrum. It is sufficient to work with positive frequency as the number of mutual impacts or meshing cycles of teeth per second.

If it is necessary to increase the distance between axis of the driving gear and the axis of the driven gear then an idler is placed between these gears that does not change the gear ratio and toothmeshing frequency. The direction of rotation of the driving and driven gears is the same.

All the basic spectrum components are usually broken down into a combination of the following effects [3].

- Low harmonics of the shaft speed.
- Harmonics of the fundamental toothmeshing frequency and their sidebands.
- Subharmonic components.
- Hunting tooth frequency components.
- Ghost (or strange) components.
- Periodicity in signals measured on planetary gearbox.
- Gear rattle.
- Components originating from faults in rolling-element bearings.

## 3.3   Low Harmonics of the Shaft Speed

Low harmonics of the shaft speed originate from unbalance, misalignments, a bent shaft, and result in low frequency vibration, therefore, without influence on the gearbox noise level due to the frequency weighting of the A-type. It concerns equipment which is connected to a gearbox via clutches, such as an engine or equipment driven via the gearbox. If the component of the rotational speed dominates the frequency spectrum of vibration, then we deal with a problem of imbalance of rotating masses. A group of low harmonics of the rotational frequency shows a parallel and angular misalignment of two shafts, which are connected to each other.

## 3.4    Harmonics of the Fundamental Toothmeshing Frequency and their Sidebands

The time course of the dynamic forces acting between the gear teeth is not a sinusoidal function of time, but ideally only a periodic function. For this reason, the frequency spectrum of the noise and vibration produced by meshing gears contains several harmonics of the fundamental toothmeshing frequency that are clearly audible because the toothmeshing and its harmonics frequency fall inside the range of human hearing. For the purpose of analysis it is usual to choose the frequency range of measurement which captures up to five harmonics of the fundamental toothmeshing frequency associated with a gear train.

An example of the instantaneous frequency spectrum is shown in Figure 3.6. An acoustic signal emitted by the transmission unit of a truck is measured during the gearbox run-up test under load. A measuring microphone is situated aside of 1 m from the contour of the gearbox in a semi-anechoic chamber. To calculate the toothmeshing frequency it is important to know the instantaneous rotational speed of any gearbox shaft, preferably during recording the block of samples for calculation of FFT. Spectrum is instantaneous without any averaging in the frequency domain. The baseband frequency range is set up to 3.2 kHz and weighting is of the A-type. The noise analysis concerns the three-stage gearbox which generally has three different sets of harmonics of the toothmeshing frequency associated with each gear train under load. The reason for this is that the gears under load generate mostly vibration and consequently noise. Only two sets of harmonics out of three dominate in the frequency spectrum of sound pressure. Their designation in the frequency spectrum in Figure 3.6 is GMF1 and GMF2. The third group of harmonics is hidden in the background of the frequency spectrum because gears rotate slowly and generated noise has a low sound level. Note the effect of A-weighting. There are no significant components below 500 Hz in the frequency spectrum.

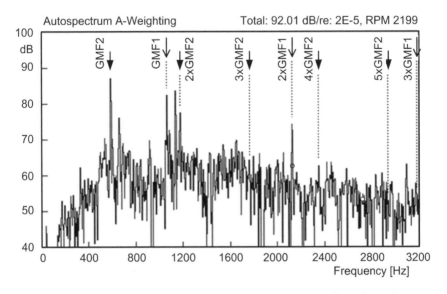

**Figure 3.6**    Noise frequency spectrum of the three-stage gearbox of a truck.

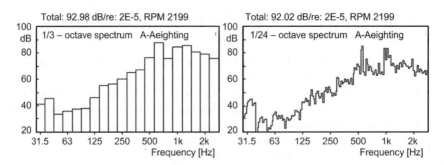

**Figure 3.7**    1/3-octave and 1/24-octave spectra of sound pressure.

There are signal analysers of the CPB type (Constant Percentage Band) which are frequently taken into account when taking sound frequency spectrum measurements. These include: the harmonics of the fundamental toothmeshing frequency and their sidebands due to the modulation effects; the noise and vibration of the geared axis systems originated from parametric self-excitation due to the time variation of tooth-contact stiffness in the mesh cycle; the inaccuracy of gears in mesh and nonuniform load and rotational speed.

Another name for these analysers is RT (Real Time) analysers or 1/3-octave analysers. In the meantime, the FFT analyser has the linear frequency scale which means that the discrete frequencies are equidistantly spaced while the previously mentioned CPB analysers have the logarithmic frequency scale. The frequency scale of the CPB analysers is characterised by the constant relative bandwidth defined as the ratio of the absolute bandwidth to the centre frequency. The RMS of the sound pressure signal is assigned to this frequency band. An example of the1/3-octave and 1/24-octave spectra is shown in Figure 3.7. As is evident from this example, the analysers of this type do not enable the individual spectrum components to be distinguished as the FFT analysers.

Operating conditions of a gear train are not steady-state. The transmitted torque by gears and rotational speed vary during rotation which causes modulation of the harmonic components of the frequency spectrum. The modulation results in rising sideband components associated with so called carrying or centre components, in our case harmonics of the toothmeshing frequency. Variation of the transmitted torque during a complete revolution causes amplitude modulation of a sinusoidal noise or vibration signal while the angular rotation speed variation causes phase modulation. The frequency of the tooth mesh cycle repeating and the frequency of rotation differ less than the carrying and modulation signals in radio transmission. In both cases it is possible to model these phenomena by modulation. Torque variation causes proportional changes of dynamic forces acting between the teeth in mesh and therefore vibrations on the surface of the gearbox housing and thus radiated noise. Amplitude and phase modulations act simultaneously.

If several gears in mesh with the different rotational frequencies have the same toothmeshing frequency, then both the sidebands belonging to the toothmeshing frequency contain the same number of sets of equally spaced components as the number of different rotational frequencies. The frequency of the carrier component can be equal to any harmonics of the toothmeshing frequency. An example of distribution of the sideband components within the carrying component is shown in Figure 3.8 which corresponds to the gear arrangement in

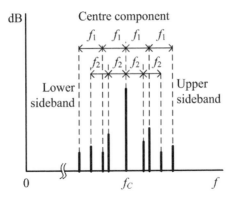

**Figure 3.8** Sidebands due to the modulation effects.

Figure 3.5. Rotational speed of gears is designated by $f_1$ and $f_2$. The frequency of the carrier component is designated by $f_C$. Differences in the amplitudes of the carrier and side components are large, so it is used as a decibel scale in Figure 3.8.

The amplitude of the components which are symmetrical about the carrier component is often different. This difference in amplitude depends on the phase shift between the amplitude modulation signal and the phase modulation signal. If both the modulation signals are in phase, then the amplitudes of the symmetrical components are identical. In the case of the phase shift between the amplitude modulation signal and phase modulation signal, that is, out of phase, the mentioned amplitudes in the sidebands are different. An example is shown in Figure 3.9. Phase modulation gives rise theoretically to an infinite number of sideband components. For simplicity, however, only one upper and one lower sideband component is shown.

An example of the vibration spectrum of a gearbox is in Figure 3.10. The centre frequency coincides with the toothmeshing frequency and the frequency range of 200 Hz is selected so as to include sideband components. The spectrum is of the ZOOM type and does not begin with 0 Hz but is spread around a selected centre frequency.

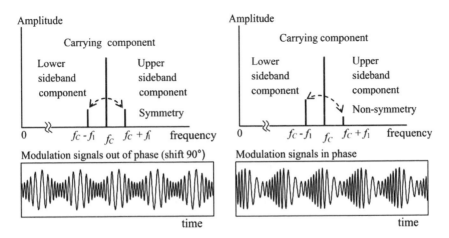

**Figure 3.9** Sideband components of the amplitude and phase modulated signal.

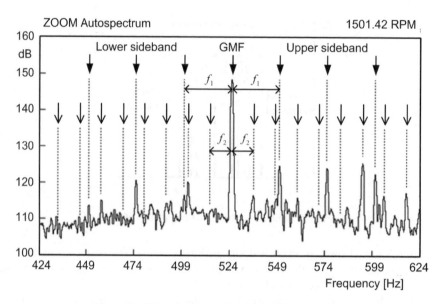

**Figure 3.10** ZOOM spectrum of a gearbox vibration.

## 3.5 Subharmonic Components

A fractional frequency of the fundamental frequency (gear rotational frequency) is called subharmonic frequency. There are nonlinear models of the tooth contact dynamics which assume the nonlinearity of tooth stiffness is a function of strain proving excitation of half of the teeth resonant frequency. This doesn't affects truck gearboxes. In the case of car gearboxes this component can be excited at such a high rotational speed that the engine noise overlays gearbox noise and therefore this phenomenon is not within the scope of this book.

### 3.5.1  Hunting Tooth Frequency

Other subharmonics originate from the rate at which the same two gear teeth mesh together. This kind of toothmesh frequency is called hunting tooth frequency $f_{HTF}$

$$f_{HTF} = f_{GMF} \frac{\gcd(n_1, n_2)}{n_1 n_2}, \tag{3.9}$$

where $\gcd(n_1, n_2)$ is the greatest common divisor of both the numbers $n_1$ and $n_2$ of teeth. [4] If the numbers $n_1$ and $n_2$ are prime then $\gcd(n_1, n_2) = 1$. The length of the repetition (hunting tooth) period between contacts of the same teeth is reciprocated by the hunting tooth frequency. In the spectrum of gearbox noise this has a negligible effect. This frequency is suitable for the control of synchronous averaging of noise and vibration signals in the time domain as shown below. When recording the length of noise and vibration signals in a steady speed gearbox it is recommended that a multiple of the length of the hunting tooth period is used.

## 3.5.2   *Effect of the Oil Film Instability in Journal Bearings*

If the machine is equipped with plain bearings then the spectrum of the vibration may, under some conditions, contain the subharmonic component with a frequency of a 0.42 to 0.48 multiple of the rotor rotational speed. The onset of the rotor instability starts when the rotational speed crosses some threshold depending on the radial clearance and oil viscosity. This phenomenon is known as the instability caused by oil film or a fluid-induced instability, commonly referred to as oil whirl.

## 3.6   Ghost (or Strange) Components

The frequency of the ghost (or strange) components is an integer multiple of the rotational speed as a toothmeshing frequency, but a gear wheel with this number of teeth is not in the gearbox. Hence the name. The ghost component of the frequency spectrum arises due to the errors in the teeth of the index-wheel of the gear cutting machine, especially the gear grinding machines which use the continuous shift grinding method. It is known that the value of the mentioned integer multiple is equal to the number of teeth of the index-wheel of this machine tool. As the number of the index-wheel teeth is large the ghost component corresponds to the pure tone of high frequency. Gear whine noise of newly built gearboxes is annoying for a customer and it is not possible to convince him that the noise will disappear over time.

The number of teeth of the index-wheel of the grinding machine can be equal to 200 or 144. It is difficult to prove that the frequency of the ghost component is such a high integer multiple of the rotational speed. The ghost component is surrounded by sideband components that may be unsymmetrical. A good method to find the multiple is to use the order analysis.

Principle of the operation of the finishing grinding machine is shown in Figure 3.11. A workpiece performs two interlinked coupled motions, rotation about its axis and translation in the direction of the workpiece axis. An error in the transfer ratio between the coupled motions of the workpiece causes regularly spaced waves on the surface of a tooth flank to be created. The number of the printed waves on the circumference of the gear is equal to the number of teeth of the index-wheel. The wave amplitude is almost immeasurable therefore these ghost components disappear after running in.

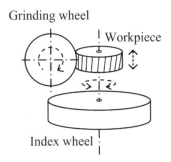

**Figure 3.11**   Continuous shift grinding method for manufacturing gears.

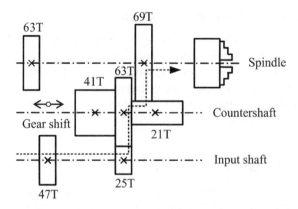

**Figure 3.12**   Lathe mechanical gearbox arrangement.

## 3.7   Gear Rattle

The above described spectrum components can be identified across all the gearbox operational conditions, except when the gear pairs are not loaded and they are vibrating within their backlash range. The unloaded gears at idle speed cause the gear rattle. It is known that the angular vibrations are dampened slightly. If there are forced angular vibrations at the driving gear of a gear train and the driven gear is connected to the lightened flywheel then a chaotic angular vibration of the driven gear occurs. The same conditions arise if there are angular vibrations of the driven gear and a heavy flywheel is driven via a gear train.

The gear rattle will be demonstrated on a lathe with a mechanical gearbox. An electric motor drives the lathe. The gearbox is connected to the motor using a V-belt. The lathe at the idle speed emits a rattling sound when the gear ratio is set up to the lowest speed of the spindle. The gearbox arrangement with the appropriate gears in mesh transfer power from the motor to the spindle rotation as shown in Figure 3.12. Changing gear ratio of the gearbox ensures the shift of three gears mounted on the countershaft in the axial direction of this shaft. It was important to find the cause of the rattle.

Gearbox vibrations were measured by an accelerometer attached to the gearbox housing and the rotational speed of the gearbox input shaft was measured by an optical tachometer probe. The probe produces a string of pulses, one pulse for each revolution of the input shaft of the gearbox. The time interval between pulses allows the average speed for one complete rotation of the input shaft to be determined. Chaotic changes of the rotation speed in RPM during a time interval of 5 seconds is shown in Figure 3.13.

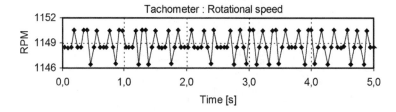

**Figure 3.13**   Time history of the rotational speed of the lathe input shaft.

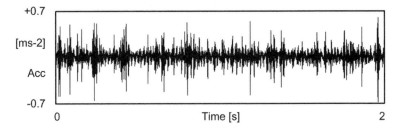

**Figure 3.14**   Time history of the gearbox housing vibration.

The record of two seconds of the vibration signal is shown in Figure 3.14. The time history of vibrations reveals rattling more clearly. The arrangement of the time history of the acceleration signal into a strip chart that is composed of the individual records corresponding to one complete revolution is analysed at end of Chapter 5.

## 3.8   Periodicity in Signals Measured on a Planetary Gearbox

An epicyclic or planetary gearbox is a gear system with planet gears revolving about a central, or sun gear which means that axes of planets revolve about the fixed axle of the central gear. Therefore the planet gears are mounted on a movable arm or carrier which itself may rotate relative to the sun gear. Calculation of the toothmeshing frequencies will consider a simple planetary gear, see Figure 3.15, in which planets are in mesh with the sun gear and with an outer ring gear or annulus which are provided by internal teeth. It is also assumed that the gearbox will reduce or drop the rotational speed of the input shaft on which the sun gear is mounted and the output shaft of the gearbox rotates at the same speed as the carrier while the outer ring does not rotate because it is attached to the gearbox housing. The numbers of teeth of the outer ring, central gear and planet are $n_1, n_2$ and $n_3$ respectively. The number of the

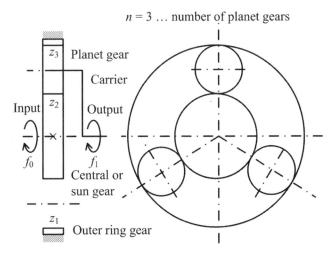

**Figure 3.15**   A simple planetary gearbox.

planet gears is designated by $n$. The rotational frequencies of the rotating parts of the gearbox are as follows:

- rotational frequency of the central (sun) gear with respect to the gearbox housing $f_0$
- rotational frequency of the carrier with respect to the gearbox housing $f_1$
- rotational frequency of the planet gear with respect to the carrier $f_P$.

The rotational frequency of the central (sun) gear with respect to the carrier is equal to $f_0 - f_1$ and the rotational frequency of the outer ring with respect to the carrier is equal to $0 - f_1$, that is, $-f_1$. Taking into account that the central gear is in mesh with the planet gears and they are in mesh with the outer ring gear via the internal teeth of this gear then the gear ratios are as follows

$$\frac{f_P}{f_0 - f_1} = -\frac{z_2}{z_3}, \quad \frac{-f_1}{f_P} = \frac{z_2}{z_3}. \tag{3.10}$$

From these two equations we can eliminate the frequency $f_P$ and get a gear ratio of the planetary drop gearbox (reducer of the rotational speed)

$$\frac{f_1}{f_0} = -\frac{z_2}{z_1 + z_2}. \tag{3.11}$$

Low frequency components or sideband components of a toothmeshing component due to the modulation effect can be found in the frequency spectrum of vibration or noise signals which are emitted by the planetary gearbox. These spectrum components are caused by a local fault of one of the tooth of an arbitrary gear when the faulty tooth enters in the mesh cycle with the other gears. The frequencies of repetition are as follows:

- meshing frequency of any tooth of the outer ring with the planet gear $f_2$
- meshing frequency of any tooth of the central gear with the planet gear $f_3$
- meshing frequency of any tooth of the planet gear with the central gear or outer ring $f_4$.

For example the meshing frequency of any tooth of the central gear with the planet gear means how many times per second the central gear enters in mesh with the planet gears. Because the selected tooth of the planetary gear enters in mesh and leaves it periodically with the sun gear and outer ring gear, a component with double frequency of $f_4$ appears in the frequency spectrum. This frequency $2f_4$ is not included in the analysis, because it is in a very simple relationship to the frequency $f_4$. The fundamental frequency $f_0$ through $f_4$ are useful for assessment of the time history of noise or vibration, which are excited by the meshing cycles of the gears. The inverse of these frequencies determines the length of the time interval for the representation of one period of the periodic signal repetition. Another set of frequencies, much higher than the frequency $f_0$ through $f_4$, is related to the frequency of the meshing cycle. A list of toothmeshing frequencies is as follows:

- toothmeshing frequency of a planet gear with the outer ring gear $f_5$
- toothmeshing frequency of all the planet gears with the outer ring gear $f_6$

- toothmeshing frequency of only one planet gear while the power flow through all the planet gears to drive the carrier is not distributed uniformly due to the tooth pitch error of the outer ring gear $f_7$.

The frequency $f_5$ assumes that the planetary gearbox is equipped with only one planet while the frequency $f_6$ is multiplied by the number $n$ of planets because the phase of the meshing cycle of the individual planets is generally shifted against each other. If all the meshing cycles are in phase then the frequency is not multiplied. The modulation effect may multiply the toothmeshing frequency of the planetary gearbox. The components with frequencies $f_5$ and $f_6$ are not always present in the spectrum.

It is known that the gears with the internal teeth of the outer ring gear are manufactured with a local error in the tooth pitch which results in a nonuniform distribution of the mechanical power flow which is transmitted by the individual planet gear train to drive the carrier. The planet gear approaching the local circular pitch error transfers full power while the other planet gears transfer reduced or even zero power. When the fully loaded planet gear passes the pair of teeth with the local circular pitch error then it becomes fully unloaded. The fully loaded operation of the planet gear corresponds to the rotation by the number $z_1/n$ of the outer ring tooth number. The number of the mesh cycles may be an integer therefore it is an integer part of the mentioned fraction of the tooth number $z_1$. Mathematically this number can be designated by an expression of $\left[z_1/n\right]$. During a complete revolution of the carrier all the planet gears become successively the most heavily loaded planet gear. Altogether there are $n\left[z_1/n\right]$ meshing cycles in total during the complete rotation of the carrier. The effect of the deviation in the circular pitch of the inner gearing in the planetary gearbox was first described in [5].

An overview of formulas to calculate all the described frequencies $f_0$ to $f_7$, using the first five fundamental frequencies $f_0$ to $f_4$ is given in Table 3.1. The fundamental frequencies $f_0$ to $f_4$ are used to control the synchronous averaging of time records.

**Table 3.1** Relationships between frequencies of all the repeating events for the planetary reducer.

| | | | | | |
|---|---|---|---|---|---|
| $f_0 =$ | $f_0$ | $\frac{z_1+z_2}{z_2}f_1$ | $\frac{1}{n}\frac{z_1+z_2}{z_2}f_2$ | $\frac{1}{n}\frac{z_1+z_2}{z_1}f_3$ | $\frac{z_3}{z_1}\frac{z_1+z_2}{z_2}f_4$ |
| $f_1 =$ | $\frac{z_2}{z_1+z_2}f_0$ | $f_1$ | $\frac{1}{n}f_2$ | $\frac{1}{n}\frac{z_2}{z_1}f_3$ | $\frac{z_3}{z_1}f_4$ |
| $f_2 =$ | $n\frac{z_2}{z_1+z_2}f_0$ | $nf_1$ | $f_2$ | $\frac{z_2}{z_1}f_3$ | $n\frac{z_3}{z_1}f_4$ |
| $f_3 =$ | $n\frac{z_1}{z_1+z_2}f_0$ | $n\frac{z_1}{z_2}f_1$ | $\frac{z_1}{z_2}f_2$ | $f_3$ | $n\frac{z_3}{z_2}f_4$ |
| $f_4 =$ | $\frac{z_1}{z_3}\frac{z_2}{z_1+z_2}f_0$ | $\frac{z_1}{z_2}f_1$ | $\frac{1}{n}\frac{z_1}{z_3}f_2$ | $\frac{1}{n}\frac{z_2}{z_3}f_3$ | $f_4$ |
| $f_5 =$ | $\frac{z_1 z_2}{z_1+z_2}f_0$ | $z_1 f_1$ | $\frac{z_1}{n}f_2$ | $\frac{z_2}{n}f_3$ | $z_3 f_4$ |
| $f_6 =$ | $n\frac{z_1 z_2}{z_1+z_2}f_0$ | $nz_1 f_1$ | $z_1 f_2$ | $z_2 f_3$ | $nz_3 f_4$ |
| $f_7 =$ | $n\left[\frac{z_1}{n}\right]\frac{z_2}{z_1+z_2}f_0$ | $n\left[\frac{z_1}{n}\right]f_1$ | $\left[\frac{z_1}{n}\right]f_2$ | $\left[\frac{z_1}{n}\right]\frac{z_2}{z_1}f_3$ | $n\left[\frac{z_1}{n}\right]\frac{z_2}{z_1}f_4$ |

$[\ldots]$ mathematical function for calculation of an integer part of the real number.

## 3.9    Spectrum Components Originating from Faults in Rolling Element Bearings

When analysing the spectrum of vibration or noise of gearboxes, we also encounter the effects of a defect of the rolling bearings; therefore the calculation of the frequency of the tonal component is part of the chapter on the spectrum signature. The gears generate a parametric vibration as a consequence of the variation of the contact stiffness of teeth in mesh during each mesh cycle. Assuming that the work surfaces of the rolling bearings are ideally flat without any defect there is no reason to excite vibration and noise. The newly produced rolling bearings, unlike the gears of the gear train, operate smoothly without any vibration due to the defects or raceways and rolling elements imperfection. The rolling element rolls on inner and outer raceways with very little rolling resistance and almost without sliding.

The nature of vibration and noise of the rolling bearings is either tonal or broadband. The defects of the rolling bearings may arise as a result of material fatigue during operation on both the inner and outer raceways and the rolling element itself. The fatigue is a consequence of shear stresses periodically appearing below the raceway surface which is exposed by load. After a period of bearing operation these stresses cause cracks which are gradually extended up to the surface. By passing the rolling elements over the cracks some fragments of material break away. This phenomenon is known as flaking or spalling. The flaking progressively increases and causes the bearing to go out of order.

There are four stages of gradual damaging of the rolling bearings. The earliest indication is in the ultrasonic frequency range. The SKF company has developed a technology called SEE$^{TM}$ Spectral Emitted Energy) for processing the signal of the 250–350 kHz frequency range excited by the metal-to-metal contact of rolling elements of bearings. The second stage is characterised by arising spectrum components outside of the audible frequency range. In the third stage the response of a pulse force which is caused by passing the rolling elements over the point where there is a defect on the raceway surface. The frequency of repeating force shocks will be computed later. The last stage of damaging causes a random broad band vibration and the rolling bearing is totally damaged.

The tonal noise and vibration of the new and healthy rolling bearings are caused by imperfections in the production process. The inner and outer rings are not as stiff as they could be and therefore they deform slightly due to the clamping for grinding. Instead of the workpiece being a circular shape the result of grinding is that the rings and raceways get an oval shape with three lobes after releasing the clamping. This excitation can be amplified by the structural resonance and therefore the harmonic components of the fundamental frequency which correspond to one fault on the raceway or rolling element appear in the frequency spectrum of noise and vibration. For the noncircularity which is composed of three lobes this frequency is an integer multiple of 3 with respect to the basic frequency. The bearings as a source of noise must be taken into account when designing a quiet gearbox.

### 3.9.1    Calculations for Bearing Defect Frequencies

When discussing the direction of the main load force rolling bearings can be divided into thrust and radial. The thrust bearings support axial load while the radial bearings support mainly radial loads. There are five types of rolling elements: a ball, a cylindrical roller, a needle, a

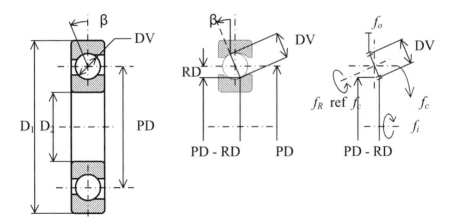

**Figure 3.16** Parameters of ball bearings.

tapered roller and a spherical roller. Calculation of the frequency of pulses excited by a defect on the rolling element and the outer and inner raceway may be generalised. The individual type of the bearing configuration and rolling element is given by a contact angle which will be explained below. We start with a description of a ball bearing in Figure 3.16. The main dimensions of the ball bearing are in Table 3.2.

Twice the distance between the bearing axis and the centre of the ball is called a pitch diameter PD. This parameter does not belong to the catalogue data, but can be estimated as the arithmetic mean of the bore and outer diameter.

$$PD = \frac{D_1 + D_2}{2} \tag{3.12}$$

The contact angle $\beta$ is defined in a plane which contains the bearing axis and the centre of the bearing ball and is measured between the radial direction and a line joining the two contact points between the ball and the inner and outer raceways (see Figure 3.17). There is an initial value of the angle $\beta$ to remove the looseness of the contact between the ball and raceway. To allow the smooth rotation of the ball bearing a slight radial looseness called a radial play is

**Table 3.2** Ball bearing dimensions.

| | |
|---|---|
| Outer diameter | $D_1$ |
| Bore | $D_2$ |
| Pitch diameter | PD |
| Ball diameter | DV |
| Contact angle | $\beta$ |
| DV cos($\beta$) | RD |
| Number of balls | $n$ |

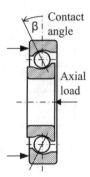

**Figure 3.17**    Contact angle of ball bearings as a result of axial load.

between the ball and the outer ring. The contact angle increases if the ball bearing is subjected to an axial load.

The contact angle of the bearing with cylindrical rollers and tapered rollers do not depend on the axial load, for the cylindrical roller the angle $\beta$ is equal to zero and for the tapered roller is the same as the angle between the bearing and roller axis as shown in Figure 3.18.

To calculate the frequencies which are excited by one defect on the raceway or roller surface we introduce the following frequencies:

- rotational frequency of the inner ring with respect to the bearing housing $f_i$
- rotational frequency of the outer ring with respect to the bearing housing $f_o$
- rotational frequency of the bearing cage with respect to the bearing housing $f_c$
- rotational frequency of the ball with respect to the bearing cage $f_R$.

The sign of frequency determines the direction of rotation. The cage frequency $f_c$ is the rotational speed of the ball or general roller cage assembly. The frequency $f_o$ is equal to zero

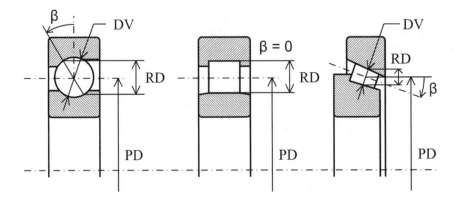

**Figure 3.18**    Contact angles of the ball, cylindrical and tapered roller.

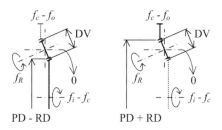

**Figure 3.19**   Two pairs of the virtual gears in mesh modelling the ball bearing.

because the bearing is mounted in the wall of the gearbox housing. There are configurations of two coaxial shafts for which both the bearing rings rotate and therefore the frequency $f_o$ is greater than zero. The calculation procedure of the frequencies which are excited by a defect is similar to the planetary gearboxes. For rolling without slide the configuration of the virtual gears is shown in Figure 3.19. The rotational frequencies of the ball and the inner and outer rings are assumed with respect to the cage; therefore the rotational frequency of the cage is equal to zero in this Figure 3.19.

The gearing ratios of two pairs of the virtual gears in mesh results from the diameters of these gears, not from the numbers of their teeth. The distance in the radial direction from the raceway on the inner ring is designated by RD and it is given by the formula

$$RD = DV \cos (\beta) \tag{3.13}$$

where DV is a diameter of the ball as a rolling element. The ratio between the rotational frequency of the ball and the rotational frequency of the inner ring with respect to the cage is as follows

$$\frac{f_R}{f_i - f_c} = -\frac{PD - RD}{DV} \tag{3.14}$$

Similarly the ratio between the rotational frequency of the outer ring and the rotational frequency of the ball with respect to the cage is as follows

$$\frac{f_c - f_o}{f_R} = -\frac{DV}{PD + RD} \tag{3.15}$$

If the Eq. (3.15) is divided by the Eq. (3.14) we get

$$\frac{f_c - f_o}{f_i - f_c} = \frac{PD - RD}{PD + RD} = \frac{1 - \dfrac{RD}{PD}}{1 + \dfrac{RD}{PD}}. \tag{3.16}$$

The ratio RD/PD may be designated by a parameter $x$

$$x = \frac{RD}{PD} = \frac{DV}{PD} \cos (\beta).$$

(3.17)

Equation (3.16) takes the form

$$\frac{f_c - f_o}{f_i - f_c} = \frac{1 - x}{1 + x}.$$

(3.18)

### 3.9.1.1  Fundamental Train Frequency

The cage frequency is named as a fundamental train frequency. This frequency is not encoun-
tered very often, but it can occur when some defect affects the rotation of the cage. The
rotational frequency of the bearing cage may be calculated from Eq. (3.18)

$$f_{FTF} = f_c = \frac{1}{2} \left[ f_o (1 + x) + f_i (1 - x) \right] = \frac{f_i + f_o}{2} - \frac{f_i - f_o}{2} x$$

(3.19)

### 3.9.1.2  Ball-Pass Frequency Outer (Race)

Passing the rolling elements over a local defect causes a pulse force which excites slightly
damped structural vibration, called ringing, at a high frequency. The vibration of the bearing
structure usually decays before the occurrence of the next pulse force. The repetitive rate of
ringing is important for the localisation of defects, that is, to decide which raceway is defected
or if there is a defect on a rolling element. The structural resonance frequency is not essential
even if the amplitude of vibration at this frequency is many times higher than the frequency
spectrum component corresponding to the frequency of the repetition ringing.

The rotational frequency of the outer ring/raceway with respect to the rotational frequency
of the cage is equal to the difference $(f_c - f_o)$, where $f_c$ is the rotational frequency of the cage
and $f_o$ is the rotational frequency of the outer ring. The Ball-Pass Frequency Outer (Race) is
the frequency of passing the balls by the defect on the outer raceway and it is designated by
$f_{BPFO}$. For $n$ balls this frequency $f_{BPFO}$ is $n$ times greater than the frequency $(f_c - f_o)$, therefore
we obtain

$$f_{BPFO} = n \left( f_c - f_o \right)$$
$$f_{BPFO} = \frac{n}{2} \left( f_i - f_o \right) (1 - x)$$

(3.20)

### 3.9.1.3  Ball-Pass Frequency Inner (Race)

The rotational frequency of the inner ring/raceway with respect to the rotational frequency of
the cage is equal to the difference $(f_i - f_c)$, where $f_o$ is the rotational frequency of the outer
ring. The Ball-Pass Frequency Inner (Race) is the frequency of passing the balls by the defect

on the inner raceway and it is designated by $f_{BPFI}$. For $n$ balls this frequency $f_{BPFI}$ is $n$ times greater than the frequency $(f_i - f_c)$, therefore we obtain

$$f_{BPFI} = n\,(f_i - f_c)$$
$$f_{BPFI} = \frac{n}{2}\,(f_i - f_o)\,(1 + x) \tag{3.21}$$

### 3.9.1.4 Ball-Spin/Roller Frequency

If there is a defect on the ball then we may assume that the ball defect impacts either only one of two raceways or both the raceways, that is, the inner and the outer raceway. If there are impacts of the ball on only one raceway then the frequency of the excited pulses, called the Ball-Spin/Roller Frequency $f_{BSF}$, is equal to the rotational frequency of the ball with respect to the rotational frequency of the cage. We obtain

$$f_{BSF} = f_R$$
$$f_{BSF} = \frac{1}{2}\,(f_i - f_o)\,\frac{(1 - x^2)}{x}\,\cos{(\beta)}. \tag{3.22}$$

Overview of the calculated frequencies is as follows

$$f_{BPFO} = \frac{n}{2}\,(f_i - f_o)\left(1 - \frac{DV}{PD}\cos{(\beta)}\right)$$
$$f_{BPFI} = \frac{n}{2}\,(f_i - f_o)\left(1 + \frac{DV}{PD}\cos{(\beta)}\right)$$
$$f_{BSF} = \frac{1}{2}\,(f_i - f_o)\,\frac{PD}{DV}\left(1 - \left(\frac{DV}{PD}\cos{(\beta)}\right)^2\right) \tag{3.23}$$
$$f_{FTF} = f_c = \frac{(f_i + f_o)}{2} - \frac{(f_i - f_o)}{2}\frac{DV}{PD}\cos{(\beta)}.$$

Because it is difficult to determine the axial force and thus the contact angle $\beta$ at the ball bearings, the parameter $x$ is estimated as follows

$$x = \frac{RD}{PD} = \frac{DV}{PD}\cos{(\beta)} \approx 0,2 \tag{3.24}$$

For rollers of the cylindrical and tapered types the contact angle is in Figure 3.18. Rolling is also not strictly without slide. The calculated frequencies are therefore only approximate and may differ from the theoretical value. It is also important to note that the frequency of bearing faults is not an integer multiple of the rotational speed.

The formulas (3.20) and (3.21) show a proportional dependence of the fault frequencies on the relative rotational speed of the inner ring with respect to the outer ring of a bearing. The multiple of the rotational frequency of the inner ring used to calculate the mentioned fault frequency can be determined experimentally for recommended operating conditions of the rolling bearings. Some manufacturers of rolling bearings give these factors for calculation of the fault frequencies as a part of the catalogue data. The source of the bearing fault can indicate the frequency cursor of special vibration analysers.

The design of the test rig for testing rolling bearings must avoid excitation of the bearings installed in a driving unit and the auxiliary bearing supporting a shaft on which the rolling bearing under test is mounted.

### 3.9.2   Envelope Frequency Analysis

As stated above the vibration signal generated by a local fault on the working surface is composed from a string of slightly damped structural vibrations. Since the structural vibrations are weakly damped, the bursts of vibration quickly disappear. The frequency of the dominating component in the frequency spectrum is the frequency of vibration during the duration of these bursts. Repetition frequency of the bursts is usually not recognisable in the frequency spectrum.

Much more relevant for assessing the frequency of repetition of the bursts in the original signal is the envelope of this time domain signal. The dominating component in the frequency spectrum of the signal envelope is the frequency of repetition.

An example showing the radial vibration signal measured on the outer ring of the tapered bearing with a local defect is in Figure 3.20. The defect was manufactured by an electrical current pulse caused by a discharge of the electrical capacitor.

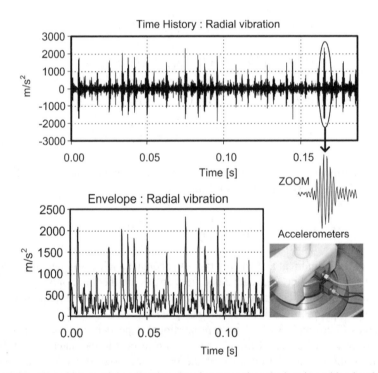

**Figure 3.20**   Time history of the vibration signal measured on the bearing with a local defect.

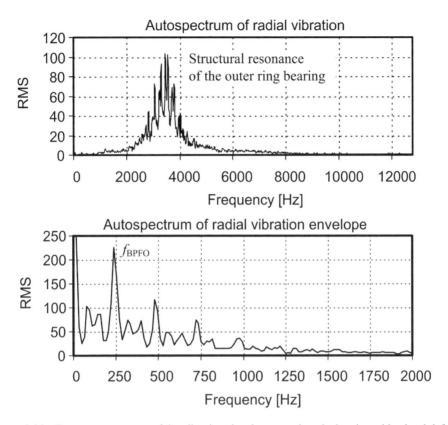

**Figure 3.21** Frequency spectrum of the vibration signal measured on the bearing with a local defect.

   The frequency spectra of the vibration signals from Figure 3.20 are shown in Figure 3.21. As already mentioned above, the frequency spectrum of the original signal contains the dominating component corresponding to structural resonances while the frequency of the dominating component in the envelope frequency spectrum is of the pulse repetition frequency. The envelope analysis and the problem of the phase demodulation is a topic of the Chapter 4.

# References

[1] Randall, R.B. and Antoni, J. (2011) Rolling element bearing diagnostics – a tutorial. *Mechanical Systems and Signal Processing*, **25**, 485–520.

[2] Byrtus, M. and Zeman, V. (2011) On modeling and vibration of gear drives influenced by nonlinear couplings. *Mechanism and Machine Theory*, **46**(3), 375–397.

[3] Tůma, J. (2009) Gearbox noise and vibration prediction and control. *International Journal of Acoustics and Vibration*, **14**(2), 99–108.

[4] Taylor, J.I. (2000) *The Gear Analysis Handbook*, Vibration Consultants, Inc., USA. ISBN 0-9640517-1-0

[5] Tůma, J. and Kuběna, R. (1996) Analýza hluku a vibrací planetového převodu včasové a frekvenční oblasti (in Czech). In: Sborník konference Inženýrská mechanika '96, Svratka.

# 4

# Harmonics and Sidebands

Noise and vibration signals generated by mechanical transmission include many sinusoidal components of various amplitude, frequency and phase, which dominate above the background noise of the frequency spectrum. Gearbox vibration and subsequent noise is caused by mesh force acting between the meshing teeth that is periodically repeated on entering into the engagement of the teeth and afterwards leaving it. The periodic time course of a dynamic force acting between the teeth over the meshing cycle is not of the sinusoidal type but much more complicated. The frequency spectrum is composed of harmonics of a fundamental frequency. It is known that the load torque of a gear train changes with the angle of rotation which produces a change in angular velocity and therefore the gears do not rotate at a constant speed. These phenomena cause modulation of sinusoidal signals. Information on how to demodulate modulated signal enables us to understand how modulation arises.

## 4.1  Harmonics

Periodic signals can be described by the Fourier series. The particular components of a periodic signal are sinusoidal functions of time, which differ from each other in frequency, amplitude and initial phase. The frequency of all these components is an integer multiple of the fundamental or basic frequency of repetition of one signal period in time. The mathematical model of a periodic signal can be formed from pairs of the phasors rotating at the same angular velocity against each other or from the sum of the cosine and sinove series. For simplicity the following formula will be shown in the form

$$x(t) = \sum_{k=1}^{K} A_k \cos\left(k\omega_0 t + \varphi_k\right), \qquad (4.1)$$

where $\omega_0 = 2\pi/T$ is a fundamental angular frequency, $A_k$ and $\varphi_k$ are an amplitude and initial phase, respectively, of the $k$-th component. The number of the harmonic components is a finite integer number $K$ due to the practical limitation of the frequency range of measurement.

*Vehicle Gearbox Noise and Vibration: Measurement, Signal Analysis, Signal Processing and Noise Reduction Measures,*
First Edition. Jiří Tůma.
© 2014 John Wiley & Sons, Ltd. Published 2014 by John Wiley & Sons, Ltd.

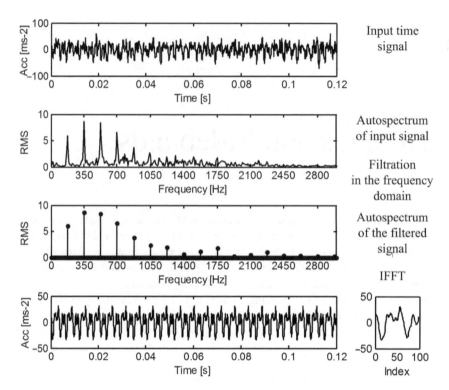

**Figure 4.1**   Calculation of the time signal from the selected harmonics.

The components of any spectrum can be considered as the harmonic components of the fundamental frequency which is equal to the reciprocal of the length of the time interval for the calculation of FFT. This section will be focused on a family of harmonics whose fundamental frequency differs from the first nonzero frequency of the spectrum. The magnitude of these harmonics in the spectrum gives only an incomplete description of the time course, which corresponds to these harmonics. A full description would consist of time course. How to calculate this time course is described in the following paragraphs.

A tool for removing frequency components other than the harmonic is through filtering in the frequency domain [1]. Then the inverse Fourier transform follows back into the time domain. The example in Figure 4.1 contains a record of 2048 samples of an acceleration signal. The spectrum of this signal contains dominant harmonic components that are superimposed by noise and therefore the original appropriate periodic signal cannot be distinguished. The fundamental harmonic component has the order of 21, another harmonic component has the order of 42 and so on. The filtering in the frequency domain restores to zero all the other components in addition to these harmonics. The result of filtering is strictly a periodic signal comprising 21 periods. One of these identical periods is shown on the right of the last diagram in Figure 4.1.

It should be noted that the number of 2048 samples is not an integer multiple of the number of 21 periods. The effect of this method of filtration is the same as the virtual splitting of the signal into 21 sections and then the calculation of the average value of the subsets of the samples from all the sections with the same index in each individual section. It is the same as averaging over time. In this example, you would have to resample the signal so that each section had the same number of samples. The filtering in the frequency domain achieves the same result without resampling.

## 4.2   Sidebands

In the modulation process we distinguish a carrier signal or component, modulating signal and modulated signal. In the description of the modulation theory, we will discuss only sinusoidal signals, which are also called harmonic signals. The carrier signal is strictly sinusoidal. The modulation signal is not limited by the generality of the sinusoidal type. For a general modulation signal the principle of superposition is valid, because the effect of each component of the modulation signal can be studied separately. An example of a harmonic time signal is as follows

$$x(t) = A \cos \left( \omega t + \varphi_0 \right) = A \cos(\Phi), \tag{4.2}$$

where $t$ is time, $A$ is an amplitude, $\Phi = \omega t + \varphi_0$ is a phase, $\varphi_0$ is an initial phase and $\omega$ is an angular frequency. Note the difference between the phase and the initial phase of the signal.

The unmodulated harmonic signal (carrier only) has constant amplitude and a phase which is a linear function of time. Modulation is the process of varying one or more properties of the previously mentioned carrier signal, with the modulating signal, which may contain information about changing the transferred power or fluctuation of the rotational speed. There are three types of modulation, that is, amplitude, phase and mixed modulation. Each type will be described separately. Effect of the modulation type on the signal amplitude and phase is shown in Figure 4.2. The parameters of the modulation signal will be defined later.

Meanwhile, it is assumed that the carrier component is a strictly harmonic signal. Generally it can also be a periodic signal whose frequency spectrum is composed of harmonics, that is, the components whose frequency is an integer multiple of the fundamental frequency. It is practical to assume that every component of this kind is modulated by the same modulating signal. Of course, this may not always be the case.

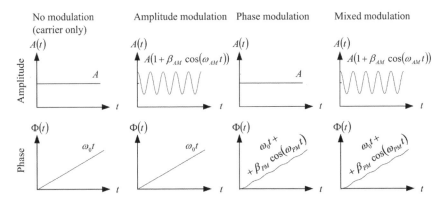

**Figure 4.2**   Effect of the modulation type on the signal amplitude and phase.

## 4.2.1   Amplitude Modulation

If it is assumed that the modulation signal is a sinusoidal signal then the amplitude modulated (AM) signal describes the following formula

$$x(t) = A \left(1 + \beta_{AM} \cos \left(\omega_{AM}t + \varphi_{AM}\right)\right) \cos \left(\omega_0 t\right), \tag{4.3}$$

where $\omega_0 = 2\pi f_0$ is a carving frequency, $\omega_{AM} = 2\pi f_{AM}$ is a modulating frequency and $\beta_{AM}$ is a modulation index. In mechanical systems the amplitude modulation can model variation of the amplitude of the noise or vibration.

Using the basic trigonometric formulas we get

$$
\begin{aligned}
x(t) &= A \left(1 + \beta_{AM} \cos \left(\omega_{AM}t + \varphi_{AM}\right)\right) \cos \left(\omega_0 t\right) \\
&= A \cos \left(\omega_p t\right) + A\beta_{AM} \cos \left(\omega_{AM}t + \varphi_{AM}\right) \cos \left(\omega_0 t\right) \\
&= A \cos \left(\omega_0 t\right) + A\beta_{AM}/2 \left(\cos \left(\left(\omega_0 - \omega_{AM}\right)t - \varphi_{AM}\right) + \cos \left(\left(\omega_0 + \omega_{AM}\right)t + \varphi_{AM}\right)\right).
\end{aligned}
\tag{4.4}
$$

The amplitude modulated signal contains three sinusoidal components. It is an original carrying component and two additional components, and they are named upper and lower sideband components according to the relation of their frequency with the carrying component frequency. One of the effects of the modulation is that the amplitude of the carrier component does not change and the amplitude of the sideband components equals to $A\beta_{AM}/2$. The frequencies of the upper $(f_0 + f_{AM})$ and lower $(f_0 - f_{AM})$ sideband components differ from the frequency $f_0$ of the carrying component and this difference equals the frequency of the modulating signal. The initial phase of the sideband components has opposite signs.

No restrictions on the relationship between the frequency of the carrier component and the frequency of the modulation signal result from Eq. (4.3). If the same modulation signal modulates a number of harmonic components, then it is appropriate for each of the sideband components to assign a carrier component which is closest in the frequency spectrum. Therefore, it holds that $f_{AM} < f_0/2$ where $f_0$ is the fundamental frequency of a set of harmonics which are associated with this fundamental frequency.

Chapter 2 has presented a model of a harmonic signal in Figure 2.18, which was created from two phasors rotating in the opposite direction

$$\cos \left(\omega t\right) = \left(\exp \left(+j\omega t\right) + \exp \left(-j\omega t\right)\right)/2. \tag{4.5}$$

Using the same procedure as in the case of the harmonic signal a model of the amplitude-modulated signal can be created as shown in Figure 4.3. The original model is extended by phasors corresponding to the sideband components of the carrier component. If the rotation of the sideband phasors is related to the rotation of the carrier phasor then the sideband phasors rotate in the opposite direction in a coordinate system which is tied with the carrier phasor

$$
\begin{aligned}
x^+(t) &= \frac{A}{2}e^{+j\omega_0 t} + \frac{A\beta_{AM}}{4}e^{+j(\omega_0 t + \omega_{AM}t + \varphi_{AM})} + \frac{A\beta_{AM}}{4}e^{+j(\omega_0 t - \omega_{AM}t - \varphi_{AM})} \\
x^-(t) &= \frac{A}{2}e^{-j\omega_0 t} + \frac{A\beta_{AM}}{4}e^{-j(\omega_0 t + \omega_{AM}t + \varphi_{AM})} + \frac{A\beta_{AM}}{4}e^{-j(\omega_0 t - \omega_{AM}t - \varphi_{AM})}
\end{aligned}
\tag{4.6}
$$

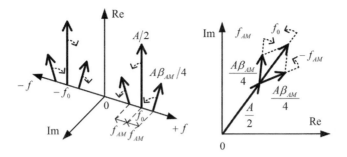

**Figure 4.3**  Phasor model of the amplitude modulation.

The sum of these two phasors extends and shortens the carrier phasor. Projection of the sum of the three phasors into the direction of the real axis results in the amplitude modulated signal as a function of time. Note that the length of the resulting phasor is the sum of the carrier component amplitude and the instantaneous magnitude of the modulation signal.

An example of an amplitude-modulated signal with a modulation index of 0.5, unit amplitude, carrier frequency of 50 Hz and a modulation frequency of 5 Hz is shown in Figure 4.4. The frequencies of all signals are chosen so that the frequency of all components matches some of the frequencies of the frequency spectrum. In such cases, the time window of the rectangular type for weighting signals is suitable. This option for the frequencies is valid for all the examples in the chapter on modulation.

Applying time windows before calculation of FFT is another example of the amplitude modulation. The use of the window of the Hanning type corresponds to the amplitude modulation. One period of the cosine function forms a modulation signal and the modulation index is equal to 1. If the frequency of the harmonic signal matches some of the frequencies of the frequency spectrum then a pair of the sideband components with half amplitude compared to the original component amplitude appears in the frequency spectrum of a weighted signal. The windows of another type have more than one pair of sideband components [2].

## 4.2.2   Phase Modulation

If it is assumed that the modulation signal is a sinusoidal signal then the phase modulated (PM) signal describes the following formula

$$x(t) = A \cos \left( \omega_0 t + \beta_{PM} \cos \left( \omega_{PM} t + \varphi_{PM} \right) \right) \tag{4.7}$$

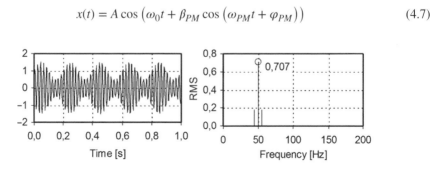

**Figure 4.4**  Amplitude modulated signal with modulation index $\beta_{AM} = 0.5$.

where $\omega_0 = 2\pi f_0$ is a carrying (or carrier) frequency, $\omega_{PM} = 2\pi f_{PM}$ is a modulating frequency and $\beta_{PM}$ is a modulation index. In mechanical systems the phase of a harmonic signal can have the same meaning as the angle of rotation. In this case the phase modulation can be used for modelling angular vibration when in rotation, in other words nonuniform rotation.

It is evident that the modulation signal, which is added to the term $\omega_0 t$, is also a harmonic signal. The instantaneous frequency of the modulated signal changes in time as follows

$$f(t) = \frac{1}{2\pi} \frac{d\Phi(t)}{dt} = \omega_0 - \beta_{PM}\omega_{PM} \sin\left(\omega_{PM}t + \varphi_{PM}\right) = f_0 - \Delta f_{PM} \sin\left(\omega_{PM}t + \varphi_{PM}\right) \quad (4.8)$$

where $\Delta f_{PM} = \beta_{PM} f_{PM}$ is called a frequency deviation. In telecommunications it is usually the carrying frequency which is much greater than the frequency deviation $f_0 \gg \Delta f_{PM}$ while for nonuniformity of rotation of a machine it is valid $f_0 > \Delta f_{PM}$.

Decomposition of the cosine function is a function of the other cosine function we will use an infinite series. The first step is to split the modulated signal into the sum of two phasors which are rotating in the opposite direction at the same speed

$$x(t) = x^+(t) + x^-(t) \quad (4.9)$$

where

$$x^+(t) = \frac{1}{2} \exp\left(+j\left(\omega_0 t + \beta_{PM} \cos\left(\Phi_{PM}\right)\right)\right) = \frac{1}{2} \exp\left(+j\omega_0 t\right) \exp\left(+j\beta_{PM} \cos\left(\Phi_{PM}\right)\right).$$

$$x^-(t) = \frac{1}{2} \exp\left(-j\left(\omega_0 t + \beta_{PM} \cos\left(\Phi_{PM}\right)\right)\right) = \frac{1}{2} \exp\left(-j\omega_0 t\right) \exp\left(-j\beta_{PM} \cos\left(\Phi_{PM}\right)\right).$$

$$(4.10)$$

where $\Phi_{PM} = \omega_{PM}t + \varphi_{PM}$ is a phase of the modulation signal.

The cosine function can be decomposed by using Euler's formula for the sum of two exponential functions and the outer exponential function can be expanded into an infinite power series. The binomial expansion of the sum of two exponential functions leads to a double sum. The result of this operation is as follows

$$\exp\left(j\beta_{PM} \cos \Phi_{PM}\right) = \sum_{n=0}^{+\infty} \frac{1}{n!} \left(\frac{j\beta_{PM}}{2}\right)^n \left(\exp\left(j\Phi_{PM}\right) + \exp\left(-j\Phi_{PM}\right)\right)^n$$

$$\exp\left(j\beta_{PM} \cos \Phi_{PM}\right) = \sum_{n=0}^{+\infty} \left(\frac{j\beta_{PM_m}}{2}\right)^n \sum_{k=0}^{n} \frac{\exp\left(j\left(n - 2k\right)\Phi_{PM}\right)}{k!\left(n - k\right)!}. \quad (4.11)$$

The following relationships $k \geq 0$, $n \geq k$ are valid for the value of the indices $k$ and $n$ in the double sum. The change of the indices by substituting $2k + i$ for $n$, the range of the new index $i$ is as follows $-\infty < i < +\infty$. A sub-part of the formula can be replaced by one of the

Bessel functions. Therefore

$$\exp\left(j\beta_{PM}\cos\Phi_{PM}\right) = \sum_{i=-\infty}^{+\infty}\sum_{k=0}^{+\infty}\left(\frac{\beta_{PM}}{2}\right)^{2k+i}\frac{(-1)^k}{k!\,(k+i)!}j^i e^{ji\Phi_{PM}} = \sum_{i=-\infty}^{+\infty}J_i\left(\beta_{PM}\right)j^i e^{ji\Phi_{PM}}$$

(4.12)

where $J_i(\beta_{PM})$ is the Bessel function of the first kind for integer orders $i = 0, 1, 2, \ldots$

$$J_i\left(\beta_{PM}\right) = \sum_{k=0}^{+\infty}\left(\frac{\beta_{PM}}{2}\right)^{2k+i}\frac{(-1)^k}{k!\,(k+i)!}.$$

(4.13)

The following identity is valid for Bessel functions with opposite indices

$$J_{-i}\left(\beta_{PM}\right) = (-1)^i J_i\left(\beta_{PM}\right).$$

(4.14)

The dependence of the Bessel function on the modulation index $\beta_{PM}$ is shown in Figure 4.5. These functions converge (tend) to zero with increasing argument, which is the modulation index $\beta_{PM}$. It should be noted that $\lim_{a\to+0} J_0(a) = 1$ and $\lim_{a\to+0} J_i(a) = 0$ for $i > 0$.

Taken into account the Eq. (4.14) the formula (4.12) can be split into two sums as follows

$$\exp\left(j\beta_{PM}\cos\Phi_{PM}\right) = J_0\left(\beta_{PM}\right) + \sum_{i=1}^{+\infty}J_i\left(\beta_{PM}\right)j^i\left(e^{ji\Phi_{PM}} + e^{-ji\Phi_{PM}}\right)$$

(4.15)

The phasor (4.10) is given by the simplified formula

$$x^+(t) = \frac{1}{2}\left(J_0\left(\beta_{PM}\right)e^{j\omega_0 t} + \sum_{i=1}^{+\infty}J_i\left(\beta_{PM}\right)j^i\left(e^{j(\omega_0 t + i\Phi_{PM})} + e^{j(\omega_0 t - i\Phi_{PM})}\right)\right).$$

(4.16)

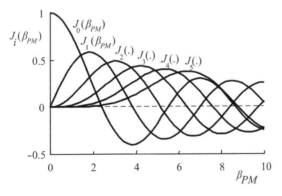

**Figure 4.5**   Bessel functions of the first kind for integer orders vs. modulation index.

The sum of the phasors rotating in the opposite direction has to be decomposed into the partial sums of even indexed terms, odd indexed terms and the term having the zero index. The final form of the formula for calculation of the modulated signal (4.9) which is composed of the harmonic functions is as follows

$$x(t) = J_0(\beta_{PM})\cos(\omega_0 t)$$

$$+(-1)^m \sum_{m=1}^{+\infty} J_{2m}(\beta_{PM})(\cos(\omega_0 t + 2m\Phi_{PM}) + \cos(\omega_0 t - 2m\Phi_{PM}))$$

$$-(-1)^m \sum_{m=0}^{+\infty} J_{2m+1}(\beta_{PM})(\sin(\omega_0 t + (2m+1)\Phi_{PM}) + \sin(\omega_0 t - (2m+1)\Phi_{PM}))$$

$$\tag{4.17}$$

where the argument of the cosine functions for $i = 1, 2, 3, \ldots$ can be rearranged in such a way

$$\omega_0 t + i\Phi_{PM} = \omega_0 t + i\left(\omega_{PM}t + \varphi_m\right) = \left(\omega_0 + i\omega_{PM}\right)t + i\varphi_{PM}$$

$$\omega_0 t - i\Phi_{PM} = \omega_0 t - i\left(\omega_{PM}t + \varphi_m\right) = \left(\omega_0 - i\omega_{PM}\right)t - i\varphi_{PM}.$$

$$\tag{4.18}$$

The structure of the formula (4.17) shows that in addition to the carrier component with the frequency $f_0$ the frequency spectrum of the phase modulated signal also contains an infinite number of sideband components whose frequencies are shifted by the frequency $f_{PM}$ of the modulation signal. The upper and lower sideband is composed of a family of harmonics of the first sideband component whose frequency is related to the frequency of the carrier component.

Because the amplitude and phase modulating signal are related to each other it is convenient to assume the same relationship between the frequency of the carrier component and the frequency of the modulation signal as in the case of the amplitude modulation. Therefore, it holds that $f_{PM} < f_0/2$. The phase modulation may also include harmonics which are connected to a fundamental frequency.

The process of phase modulation can be represented by adding the phasors together. This addition is shown in Figure 4.6 in a 3D diagram. Only the phasors associated with a pair of the first sideband components of the carrier component are shown on the left side of the

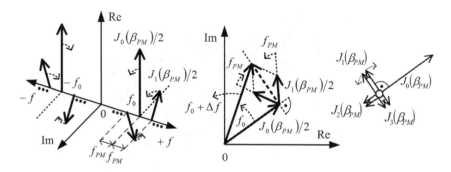

**Figure 4.6**   Phasor model of the phase modulation.

figure. If the rotation of the sideband phasors is related to the rotation of the carrier phasor then the sideband phasors rotate in the opposite direction in a coordinate system which is connected to the carrier phasor. Other phasors are omitted for clarity. The sum of the phasors belonging to a pair of the first sideband components is shown in the centre of the figure. Unlike amplitude modulation the resulting phasor is perpendicular to the phasor belonging to the carrier component, that is, it rotates by a quarter of a turn in a clockwise direction. The result of adding all sideband components together is shown in the diagram on the right of Figure 4.6. Multiplication of a phasor by the imaginary unit which is taken to the positive integer power $j^i$ causes a phase shift forwards in a clockwise direction by the multiple of a quarter turn which equals to the exponent of the imaginary unit. Projection of the phasor into the direction of the real axis results in the phase modulated signal as a function of time. Note that the rotation angle of the resulting phasor is the sum of the rotation angle of the carrier component and the instantaneous phase modulation signal which is measured from the position of the carrier phasor.

An example of a phase modulated signal is shown in Figure 4.7. The carrying component has a frequency of 50 Hz while the phase modulation signal has a frequency of 5 Hz. Because the modulation index is equal to 5 then the frequency deviation is 25 Hz. This means that the instantaneous frequency ranges from 25 to 75 Hz. The components of the spectrum with significant amplitude belong to the frequency range from 0 to 100 Hz. The carrying component does not dominate in the spectrum $f_0 \gg \Delta f_{PM}$.

Several examples documenting the effect of the value of the modulation index on the frequency spectrum appearance of the phase-modulated signal are shown in Figure 4.8. The frequency range of the spectrum is from 0 to 3200 Hz. The carrying component has a frequency of 1600 Hz, which is situated in the middle of the spectrum frequency range.

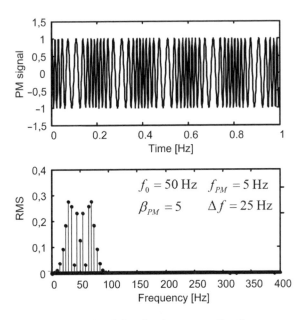

**Figure 4.7**   Phase modulated signal and corresponding frequency spectrum.

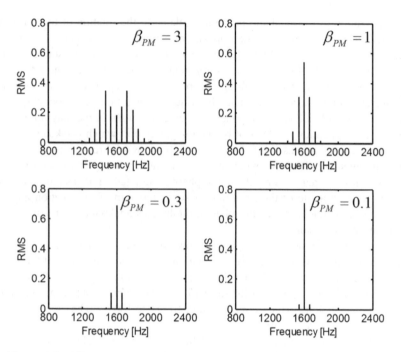

**Figure 4.8**  Effect of the modulation index on the RMS of the sideband amplitude.

To achieve high resolution spectra only half of the range is displayed and the centre of the zoomed frequency range is placed in the carrier frequency.

Indoor tests of noise and vibration of transmission units take place with a continuous increase of the input rotational speed. This test simulates the acceleration of a vehicle for measurement of outdoor pass-by noise on a test track. Another reason why they do not test a transmission unit of a vehicle at a constant rotational speed is that the nominal RPM are not specified and the engine operates in a very wide rpm range, such as trucks up to 2400 RPM and passenger cars up to 5000 RPM and above. During the development of a new product or quality control of the production it is necessary to examine the entire operating speed range for various operational conditions to eliminate disturbing noise or vibration. These tests are carried out over run up or coast down of the gearbox.

As stated earlier it is necessary to create a sequence (record) of signal samples from the time interval of a certain length $T$ for calculating the frequency spectrum by using Fourier transform. The frequency of the dominant components in the spectrum varies in accordance with increasing or decreasing the rotational speed of an input shaft. The change of the frequency in the input signal for calculation of the spectrum can be regarded as a frequency modulation. The effect of this modulation will be demonstrated with the use of an example.

The relative change of the frequency over an interval of the length $T$ in percentage terms can be defined as follows

$$p = \frac{|f_T - f_{REF}|}{f_{REF}} 100\%$$

(4.19)

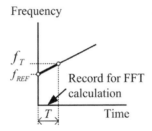

**Figure 4.9**   Increase of the signal frequency during measurements.

where $f_{REF}$ is the frequency at the beginning and $f_T$ is the frequency at the end of the time interval of the length of $T$. The parameters in the formula (4.19) are illustrated in Figure 4.9.

To demonstrate the effect of the frequency modulation according to a linear function of time the frequency spectrum of the sinusoidal signal is calculated for various rates of increasing the frequency of this signal. The reference signal whose amplitude is equal to unity has an initial frequency of 1 kHz. The sampling frequency of this testing signal is equal to 8192 Hz which results in a frequency spectrum ranging from 0 to 3200 Hz. The time record for calculation of 800 spectral lines contains 2048 samples and has a length of 250 ms, that is, a quarter of a second. The difference between adjacent components for this time interval is 4 Hz. The time window of the Hanning type is used to weight or window the time signal. The signal frequency is increased at different rates from 0.1 to 5% per record. The percentage figure determines how much the initial frequency increases in hertz over the mentioned time interval. Results of the calculation are shown in Figure 4.10. The spectrum of the signal which has constant (stationary) frequency contains only three components, that is, the component of 1 kHz and two sideband components $1600 \pm 4$ Hz resulting from the use of the Hanning time window. These three nonzero components form a spectral peak. If the frequency of the tracked component varies in the time interval corresponding to a segment of samples for calculation of FFT then the number of nonzero components forming the peak increases while their amplitude simultaneously decreases. This phenomenon can be designated as smearing the original narrow and high spectral peak. In other words such a peak appears to be smeared compared to the original peak and its power is spread across a wider frequency band. This conclusion follows from a comparison of the spectrum for $p = 0\%$ and $p = 5\%$ in Figure 4.10. The average centre frequency of 1024 Hz corresponds to the rate of the frequency increase by 5%. The smeared peak is composed from the nonzero components whose frequency ranges approximately from 1000 to 1050 Hz. The simulation shows that the relative bandwidth corresponds to the relative change in frequency over the time interval of 250 ms.

The more a tracked component of the spectrum is smeared, the lower the amplitude of the components which form the peak. With the exception of the order spectra, where smearing is excluded, it is not possible to consider a single component of the spectrum as a representative component of the peak. The value of RMS for the tracked peak in the frequency spectrum is calculated by summing the squares of the component RMS from the frequency band that is centred on the representative frequency of this peak. Then the sum of squares is divided by the time window bandwidth. The calculation of the representative RMS is completed by the calculation of the square root of the previous result.

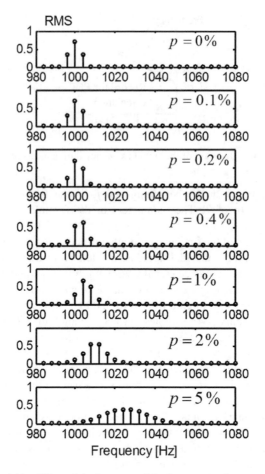

**Figure 4.10**   Effect of the increase of the frequency when measuring.

The choice of the relative bandwidth for calculating the representative RMS depends on the relative change of the rotational speed, for example, a gearbox under test. Both the relative percentages are numerically identical. If the relative bandwidth is 0.5% of a given frequency then the allowed change of RPM is also 0.5% over a time interval of measuring the time record for FFT calculation. The frequency range goes up to 3200 Hz, the relative bandwidth of 0.5% and the frequency spectrum of 800 lines needs at least 40 seconds for doubling RPM.

### 4.2.3   Mixed Modulation

In the case of mixed modulation it is also assumed that both the modulation signals are sinusoidal signals. The signal whose amplitude and phase is modulated simultaneously is described by the following formula

$$x(t) = A \left( 1 + \beta_{AM} \cos \left( \omega_{AM} t + \varphi_{AM} \right) \right) \cos \left( \omega_0 t + \beta_{PM} \cos \left( \omega_{PM} t + \varphi_{PM} \right) \right) \quad (4.20)$$

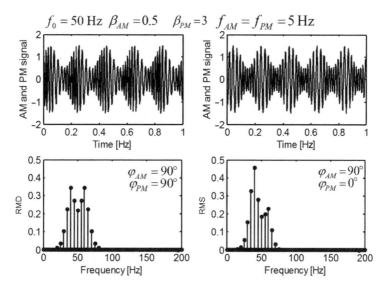

**Figure 4.11**  Amplitude and phase modulated signals.

where all the parameters were defined before. Arbitrary values of the parameters can be used in the formula for calculating signal samples. Their choice has only a practical limitation, those described above. Also of interest is the case when the two modulation frequencies are identical.

Two examples of the mixed modulation of a sinusoidal signal are shown in Figure 4.11. The frequencies for modulation of the amplitude and phase are identical for both examples. It means that the carrier component and associated sideband components are situated in the same frequencies on the frequency axis. The parameters of the modulation process are the same except for the relationship between the phases of the modulation signals for the amplitude and phase modulation. Of importance is the phase difference of the two modulation signals, one of which is for amplitude modulation and the second one for phase modulation. The two examples differ in the phase difference. These phases are the same for the modulated signal on the left side and differ for the modulated signal on the right side. In the second case the phase difference is equal to $\pi/2$ radians. The symmetry of the sideband components around the carrier component in the frequency spectrum of the modulated signal is affected by the relation between the phases of the modulation signals. The zero phase shift between the mentioned phases reflects symmetry of the amplitudes of the sideband components while the modulation signals which are out of phase have resulted in a nonsymmetry of the amplitude of the sideband components in the frequency spectrum of the modulated signal. The modulation signals in phase have lead to the symmetry around the carrier component while the modulation signals with a phase shift result in nonsymmetry of the amplitude of the sideband components.

## 4.3  Analytic Signal

Knowledge of the position and length of the phasor rotating in the positive direction for the modulated signal enables modulation signals to be easily isolated and their amplitude,

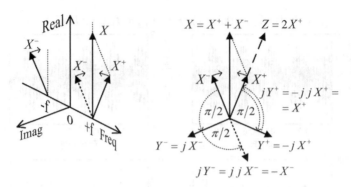

**Figure 4.12** Calculation of the analytic signal.

frequency and initial phase to be calculated. A phasor rotating in the positive direction is called the analytic signal. The problem that needs to be overcome is how to delete the phasor which rotates in the negative direction. The calculation procedure is shown in Figure 4.12. The solution will be demonstrated by the harmonic signal that results from the sum of two phasors $X^+$ and $X^-$ rotating against each other at the same speed of rotation. This procedure is sufficient because each periodic signal can be decomposed into the sum of harmonics [2]. The sum $X$ of the phasors which creates a real number is designated as follows

$$X = X^+ + X^- \tag{4.21}$$

Although the phasors are a complex number, their sum is a real number, which varies over time. It is a time real-valued function or signal $x(t)$. To delete the phasor $X^-$ a phasor $-X^-$ which rotates in the opposite phase and in the same direction is needed. The two new phasors $Y^+$ and $Y^-$ corresponding to a real-valued function or signal $y(t)$ will be created in the first instance. The phase of the phasor rotating in the positive direction is shifted by a quarter of a turn back, that is, in a negative direction. This phase shift corresponds to the multiplication of the phasor $Y^+$ and a negative imaginary unit. The phase of the phasor rotating in the negative direction is shifted by a quarter of a turn ahead, that is, in the positive direction. In contrast to the previous multiplication of the phasor $Y^-$ is multiplied by a positive imaginary unit

$$Y^+ = -jX^+, \quad Y^- = jX^-. \tag{4.22}$$

The last formulas define the Hilbert transform. A graphical representation of mathematical operations is shown in Figure 4.12. The sum of the phasors, which have been calculated by the formula (4.22) is as follows

$$Y = Y^+ + Y^- = -jX^+ + jX^- = -j\left(X^+ - X^-\right). \tag{4.23}$$

The analytic signal is a complex signal with an imaginary part, which is the Hilbert transform of the signal real part. The definition of the analytic signal $Z$ is given by the following formula

$$Z = X + jY = X^+ + X^- + j\left(Y^+ + Y^-\right) = X^+ + X^- + \left(X^+ - X^-\right) = 2X^+. \tag{4.24}$$

The imaginary part of the definition formula for the analytic signal was replaced by the result of the calculation of the formula (4.23) which shows how the Hibert transform $Y$ depends on the phasors of the original signal $X$. The definition of the analytic signal indicates that this is a double the phasor of the initial signal that rotates in the positive direction. Doubling the phasor can be explained by the fact that the magnitude of the phasor is half the amplitude of the real signal as can be seen in Figure 4.12.

### 4.3.1   Definition of Hilbert Transform

In this chapter, the Hilbert transform is defined first for continuous-time signals, which contain sinusoidal components in the frequency spectrum. The signals having this property are periodic signals. The Hilbert transform changes the signal from the time domain back again into the time domain as a filter, and therefore has the properties of the filter whose characteristics can be described by a transfer function, which shows the relationship between the Fourier transforms of the continuous input and output signal. The filter, which acts as the Hilbert transform is called a Hilbert transformer or a 90-degree phase shifter. Let $X(j\omega)$ and $Y(j\omega)$ be the Fourier transform of the input signal $x(t)$ and the output signal $y(t)$, respectively. From formulas (4.22) it follows that the Hilbert transformer shifts the phase by $\pi/2$ radians

$$H_{HT}(j\omega) = \frac{Y(j\omega)}{X(j\omega)} = \begin{cases} -j, & \omega > 0 \\ +j, & \omega < 0. \end{cases} \tag{4.25}$$

The phase shift causes the sine function to become the cosine function after transforming , and vice versa. Figure 4.13 illustrates this.

The impulse response of the Hilbert transformer also describes the filter properties in addition to the transfer function. The impulse response is calculated from the frequency response by using the inverse Fourier transform. The direct calculation of the inverse transformation of the transfer function (4.25) is not possible, because this function is nondecaying. To convert the nondecaying function to the decaying function, the frequency transfer function has to be extended by an exponential function which is easily inferable. The original transfer function can be obtained if the variable $\sigma$ tends to zero as shown in Figure 4.14.

$$H(j\omega) = \begin{cases} -je^{-\sigma\omega}, & \omega > 0 \\ +je^{+\sigma\omega}, & \omega < 0 \end{cases} \qquad \lim_{\sigma \to 0} H(j\omega) = H_{HT}(j\omega). \tag{4.26}$$

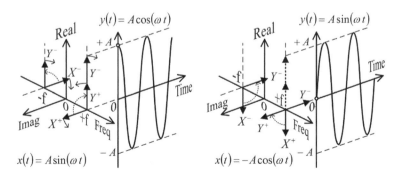

**Figure 4.13**   Hilbert transform of the sine and cosine function.

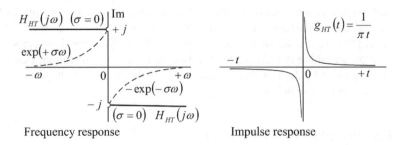

Frequency response                                  Impulse response

**Figure 4.14**   Frequency and impulse response of the Hilbert transform.

The impulse response is first calculated for a modified transfer function $H(j\omega)$ as follows

$$g(t) = \frac{1}{2\pi} \int_{-\infty}^{+\infty} H(j\omega)\, e^{j\omega t}\, d\omega = \frac{1}{2\pi} \left[ j \int_{-\infty}^{0} e^{\sigma\omega + j\omega t}\, d\omega - j \int_{0}^{+\infty} e^{-\sigma\omega + j\omega t}\, d\omega \right]$$

$$= \frac{j}{2\pi} \left( \left[ \frac{e^{\sigma\omega + j\omega t}}{\sigma + jt} \right]_{-\infty}^{-0} - \left[ \frac{e^{-\sigma\omega + j\omega t}}{-\sigma + jt} \right]_{+0}^{+\infty} \right) = \frac{t}{\pi\left(\sigma^2 + t^2\right)}, \tag{4.27}$$

We get the impulse response of the Hilbert transformer after calculation of the following limit

$$g_{HT}(t) = \lim_{\sigma \to 0} g(t) = \lim_{\sigma \to 0} \frac{t}{\pi\left(\sigma^2 + t^2\right)} = \frac{1}{\pi t}. \tag{4.28}$$

The impulse response of the Hilbert transformer in Figure 4.14 is not a causal filter and is not defined for $t = 0$. A response precedes the change at the system input. This filter is theoretically realisable; its properties can only be approximated.

The Hilbert transform of a time function can be calculated similarly to the Fourier transform by a definition formula. Because it concerns a filter the convolution integral is the default formula for this definition. A special feature of this definition is the use of the Cauchy principal value (PV) which expands the class of certain improper integrals for which the finite integral exists as, for example, the integral

$$\lim_{\varepsilon \to 0+} \left[ \int_{a}^{\xi - \varepsilon} f(x)\, dx + \int_{\xi + \varepsilon}^{b} f(x)\, dx \right] \tag{4.29}$$

where

$$\xi \in (a, b), \quad \int_{a}^{\xi} f(x)\, dx = \pm\infty, \quad \int_{\xi}^{b} f(x)\, dx = \mp\infty. \tag{4.30}$$

**Table 4.1**   Hilbert transform.

| Signal | Hilbert transform |
|---|---|
| $\sin(\omega t)$ | $-\cos(\omega t)$ |
| $\cos(\omega t)$ | $\sin(\omega t)$ |
| $1/(t^2 + 1)$ | $t/(t^2 + 1)$ |
| $\sin(t)/t$ | $(1 - \cos(t))/t$ |
| $\delta(t)$ | $1/\pi t$ |

The Hilbert transform of a function of $x(t)$ into a function $y(t)$ can be defined as the principal value integral

$$y(t) = \frac{1}{\pi} \text{P.V.} \int_{-\infty}^{+\infty} \frac{x(\tau)}{t - \tau} d\tau \qquad (4.31)$$

Some examples of the transformation are shown in Table 4.1.

### 4.3.2   Calculation of Hilbert Transform of Sampled Signals

Practical calculations can be realised only with sampled signals. The calculation of the Hilbert transform of any sampled signal can only be done using the numerical procedure for the input data in the form of the equidistantly sampled signal $x_n$, $n = 0, 1, 2, \ldots$ and the index $n$, called the sample number, may range over some or all the integers. Two methods are used:

- Fast Fourier Transform (FFT).
- Digital filters.

Digital filters are useful for continuous calculations. Their disadvantage is a phase delay. In this book, the input data corresponds to records of sampled signals over an appropriate period of time, for example, the time interval of the complete revolution of a machine and therefore attention will first be focused on the use of FFT. It is assumed that the number of the samples in the record is an even number $N$. Calculation with the use of FFT is without any phase delay in the output signals.

The algorithm for computing the Hilbert transform of the samples $x_n$, $n = 0, 1, 2, \ldots, N - 1$ with the use of FFT is broken down into three steps:

A. The Fast Fourier Transform (FFT) of the input signal which contains a limited number of samples to obtain phasors $X_k$, $k = 0, 1, 2, \ldots, N - 1$, rotating in positive and negative direction,

$$
\begin{aligned}
X_k^+ &= X_k, \quad k = 1, 2, \ldots, N/2 - 1, \\
X_{N-k}^- &= X_k, \quad k = N/2 + 1, \ldots, N - 1.
\end{aligned}
\qquad (4.32)
$$

B.  Rotation of the phasors $X_k^-$ in the positive direction by the angle of $+\pi/2$ radians and the phasors $X_k^+$ in the negative direction by the angle of $-\pi/2$ radians

$$Y_k^+ = -jX_{P,k} = -j\left(\text{Re}\left(X_k^+\right) + j\text{Im}\left(X_k^+\right)\right) = \text{Im}\left(X_k^+\right) - j\text{Re}\left(X_k^+\right)$$
$$Y_k^- = +jX_k^- = j\left(\text{Re}\left(X_k^-\right) + j\text{Im}\left(X_k^-\right)\right) = -\text{Im}\left(X_k^-\right) + j\text{Re}\left(X_k^-\right).$$

(4.33)

Exchanging the real and imaginary parts causes a rotation by a quarter of a turn.

C.  Preparation of data for calculation of the inverse fast Fourier transform (IFFT)

$$Y_0 = Y_{N/2} = 0$$
$$k = 1, 2, \ldots, N/2 - 1:$$
$$Y_k = Y_k^+$$
$$Y_{N-k} = Y_k^-.$$

(4.34)

The last step of the calculation is the inverse Fourier transform of the complex data $Y_k$, $k = 0, 1, 2, \ldots, N - 1$. The result is the signal $y_n$, $n = 0, 1, 2, \ldots, N - 1$.

The calculation of the Hilbert transform is often associated with filtering in the frequency domain. Filtration concerns the imaginary and real part of the analytic signal, which then forms a sequence of complex samples.

The calculation of the Hilbert transform may be done with the use of digital filters. There are two types of filters, namely FIR and IIR. The preferred filters are of the FIR type due to the linearity of the phase which only causes a delay of the signal at the output of the filter and is free of distortion. It is possible to design a suitable filter of the IIR type if the order of the filter has to be small.

The transfer function of a digital version of the Hilbert transformer is the following:

$$G_{HT}\left(j\omega\right) = \frac{Y\left(j\omega\right)}{X\left(j\omega\right)} = \begin{cases} -j, & +\pi > \omega T_S > 0 \\ j, & -\pi < \omega T_S < 0. \end{cases}$$

(4.35)

This transfer function is different from the transfer function of signals with continuous time in the range of angular frequencies. The impulse response is given by the formula

$$g_{HT,n} = \frac{1}{2\pi} \int_{-\pi}^{+\pi} G_{HT}\left(e^{j\omega T_S}\right) e^{j\omega T_S n} \,\mathrm{d}\left(\omega T_S\right) = \begin{cases} 0, & n = 2k \\ 2/\pi n, & n = 2k + 1. \end{cases}$$

(4.36)

The individual values of the impulse response in Figure 4.15 with an even index are zero and the odd index decreases proportionally to the inverse of the index value. The result of the calculation is the impulse response of the ideal Hilbert transformer, which is unrealisable.

The impulse response of a filter is a sequence of the nonzero samples $y_n$ at the output of this filter when the sequence of samples at the input $x_n$ of the filter contains only one nonzero

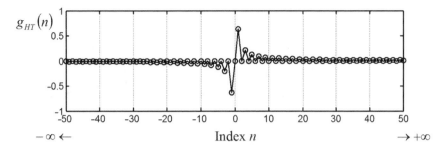

**Figure 4.15** Impulse response of the ideal Hilbert transformer.

sample. The coefficients $b_0, b_1, \ldots, b_M$ of the FIR filter are identical to the nonzero samples of the filter impulse response. The output samples of the FIR filter of the order $M$ are calculated according to the formula

$$y_n = b_0 x_n + b_1 x_{n-1} + \cdots + b_{M-1} x_{n-M+1} + b_M x_{n-M}, \quad n = 0, 1, 2, \ldots \qquad (4.37)$$

The impulse response of the ideal Hilbert transformer can be symmetrically reduced to a finite number of nonzero samples while preserving anti-symmetric coefficients. The causality of the FIR filter is ensured by a shift of $M/2$ samples in the right side direction of the time axis.

The digital filter acts as a Hilbert transformer only for a frequency band in which the magnitude of the frequency response function is equal to the unit. The impulse response which is corrected with the use of the Kaiser window smoothes the frequency response function.

The frequency properties of the Hilbert transformer of two versions designed for a cut-off normalised frequency 0.95 are shown in Figure 4.16. The order of both filters is 161. One of the transformers is ideal while the impulse response of the second one is weighted using the

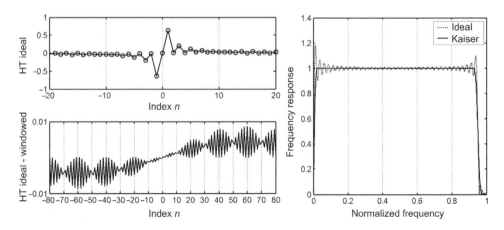

**Figure 4.16** Frequency response of the Hilbert transformer with the normalised cut-off frequency of 0.95.

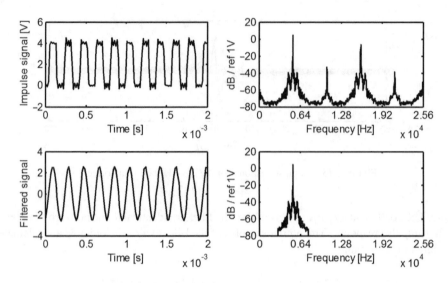

**Figure 4.17**    Bandpass filtration of the pulse signal.

Kaiser window. The difference between the two impulse responses is shown in the panel on the bottom left of the figure. The frequency responses are given in the diagram on the right. The ideal Hilbert transformer does not work properly for low and high frequency while the frequency response of the second transformer is smooth.

### 4.3.3    Demodulation of the Modulated Harmonic Signal

The theory for demodulation of the modulated signals will be described for continuous-time signals. The calculation procedure formally described for continuous time can be achieved over time. Variables that are functions of time are replaced by variables with an index, which has the meaning of discrete time.

The frequency spectrum of vibration and noise signals which are produced by the machine operation contain many spectral dominant components. The amplitude and phase of each component can be modulated by varying torque or speed of rotation. In the case of gearboxes the structure of the shafts and gears has a large mechanical stiffness and low damping. In particular, damping of torsional vibration is very small. An analysis tool for vibration and noise of the machines is demodulation of the modulated carrier components and is focused on discovering sources of vibration [3].

The demodulation method can handle only a single modulated sinusoidal signal. Other parts of the frequency spectrum must be filtered out using the band pass filter. Nothing other than the carrier component and adjacent sideband components in the spectrum may remain. The bandpass filtering in the frequency domain is preferably used for conditioning the signals to be demodulated. An example of this procedure is shown in Figure 4.17. The IRC sensor produces a string of pulses which contain information about the variations of the instantaneous frequency of pulses with the use of the phase demodulation. The frequency spectrum of the pulse signal contains several harmonics of the base frequency to be analysed. The bandpass filtering is used to separate a frequency band containing the carrier component and adjacent sidebands. The total bandwidth is equal to the frequency of the carrier component. The modulated harmonic signal is obtained with the use of the inverse Fourier transform.

The sidebands of the carrier component contain information about the modulating signal. The continuous-time signal after the bandpass filtration is designated by $x(t)$. The Hilbert transform of the signal $x(t)$ is calculated after bandpass filtering. The resulting continuous-time signal is designated by $y(t)$. The continuous-time analytic signal is composed as follows

$$z(t) = x(t) + jy(t). \tag{4.38}$$

The amplitude modulation signal, referred to as an envelope, is calculated as the absolute value (or modulus) of the analytic signal with the use of the Pythagorean theorem as follows

$$A(t) = |z(t)| = \sqrt{x(t)^2 + y(t)^2} \tag{4.39}$$

The envelope is identical to twice the absolute value of the phasor which rotates in the positive direction.

The principal value of the instantaneous phase of the analytic signal is as follows

$$\Phi_{P.V.}(t) = \text{Arg}\,(z(t)) = \arctan\,(y(t)/x(t)) \tag{4.40}$$

The principal value of the instantaneous phase belongs to either the $(-\pi, +\pi)$ or $\langle 0, +2\pi)$ interval. The former is used in this book. The instantaneous phase is also called the wrapped phase. The phase in radians can be computed by the previous formula. The sign of the value of $x(t)$ and $y(t)$ is respected for assigning a quadrant in the complex plane. The result of calculation of the angle is in the wrapped form which limits the angle into the interval

$$-\pi < \Phi_{P.V.}(t) \le +\pi \tag{4.41}$$

As a last step of the phase demodulation process the wrapped phase has to be unwrapped into

$$\arg\,(z(t)) = \text{Arg}\,(z(t)) + 2\pi n(t) \tag{4.42}$$

where $n(t)$ is a sequence of integer numbers, which depends on time $t$, for that the function $\arg(z(t))$ is without discontinuities greater than a permissible value.

## 4.3.4   Unwrapping Phase

The practically realisable process of the unwrapping phase only concerns sampled signals. The algorithm for unwrapping a phase of the analytic signal is based on the rule determining the maximum possible change in phase. The value of the permitted changes in phase is related to the sampling theorem. It is assumed that a continuous harmonic signal is sampled

$$x_n = \cos\,(\omega n T_S) = \cos\,(2\pi f n T_S)\,, \quad n = 0, 1, 2, \ldots. \tag{4.43}$$

The phase of the mentioned harmonic signal is as follows

$$\Phi_n = \omega n T_S = 2\pi f n T_S. \tag{4.44}$$

The phase difference between two consecutive samples in time $t_n$ and $t_{n+1}$ where $T_S = t_{n+1} - t_n$ may be written in the form

$$\Delta\Phi = \Phi_n - \Phi_{n-1} = \omega T_S = 2\pi f T_S = \frac{2\pi f}{f_S}. \tag{4.45}$$

The Nyquist-Shannon sampling theorem requires that the sampling frequency has to be greater than the double frequency of the sinusoidal signal

$$2f \le f_S \Rightarrow \frac{f}{f_S} \le \frac{1}{2} \Rightarrow \frac{2\pi f}{f_S} \le \frac{2\pi}{2} = \pi \tag{4.46}$$

This relation shows a limited value of changes in the signal phase

$$\Delta\Phi = \frac{2\pi f}{f_S} \le \pi. \tag{4.47}$$

It can be concluded that the change in phase within the sampling interval has to be less then $\pi$ radians $|\Delta\Phi| \le \pi$. The algorithm of the phase unwrapping is based on the allowed change in phase within the sampling interval [3].

The demodulation process will be described in the example in Figure 4.18. The signal to be demodulated has a frequency of 40 Hz, the modulation signal has a frequency of 2 Hz and a modulation index equals to 5. The sampling frequency is 256 Hz.

$$x(t) = \sin(2\pi 20t + 5\sin(2\pi 2t)) \tag{4.48}$$

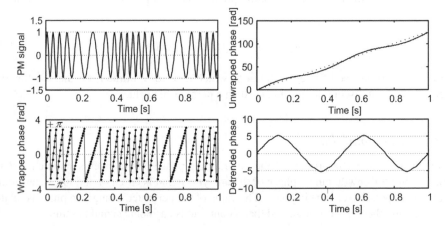

**Figure 4.18**    Phase demodulation.

The first step of the phase demodulation process is to calculate the wrapped phase which is of the sawtooth wave shape as can be seen in bottom, left hand panel in Figure 4.18. As is evident from the figure, there are discontinuities in the sequence of the magnitude of the instantaneous phases that can be removed by using the algorithm of unwrapping which converts discontinuities in instantaneous phase as a function of time to a continuous function of time. The algorithm of unwrapping monitors the change $\Delta\Phi_{P.V.}$ of the consecutive samples of the instantaneous phase. If a change is detected greater than the angle of $\pi$ radians, then all the further samples of the phase will be shifted by $\pm\pi$ so that the jumps in the unwrapped phase samples are removed. An integer $n$ in the unwrapping algorithm counts the number of preceding step changes $\pm\pi$ in phase. The initial value of $n$ is equal to zero

$$
\begin{aligned}
\Delta\Phi_{P.V.} &\leq -\pi & &\Rightarrow & 2\pi n + \Phi_{P.V.} + 2\pi &\rightarrow \Phi, n+1 \rightarrow n \\
-\pi < \Delta\Phi_{P.V.} &< +\pi & &\Rightarrow & 2\pi n + \Phi_{P.V.} &\rightarrow \Phi \\
\Delta\Phi_{P.V.} &\geq +\pi & &\Rightarrow & 2\pi n + \Phi_{P.V.} - 2\pi &\rightarrow \Phi, n-1 \rightarrow n.
\end{aligned}
\tag{4.49}
$$

The unwrapped phase on the upper panel on the right side of Figure 4.18 consists of a phase corresponding to a constant frequency (dotted line) of the carrier component and a superposed part (solid line) which is a modulation signal with the amplitude of five radians.

The phase modulation signal is the difference between the unwrapped phase and the linear ramp $(\omega_0 t)$. This calculation removes the linear trend from the phase as a function of time and, therefore, the result can be considered as a detrended phase. This calculation terminates the process of the phase demodulation.

### 4.3.5  Normalising Phase

The phase demodulation of a pulse signal is used to analyse the variation of the machine rotational speed. The source of pulses may be an IRC sensor which is attached to the rotating shaft. The strings of pulses which are generated by a built-in sensor for ECU (engine control unit) are also applicable to measuring the variability of the engine rotation. This sensor indicates the passing of the pins on the flywheel. The ideal source of the pulse signal is acceleration or the noise of a gearbox. The meshing gear produces a harmonic signal whose amplitude and phase are modulated. In this case, it is necessary to use a bandpass filtering system to separate a string of pulses from other components of the spectrum.

Let $K$ be the number of pulses generated by a sensor per revolution. This number also corresponds to the number of periods of the modulated harmonic signal, which originates from the pulse signal after filtration. The overall change in the unwrapped phase of the harmonic signal for one complete revolution is equal to $+2\pi K$. This phase change depends on the number of the pulses. For analysis of the uniformity of rotation it is necessary to relate this change in phase to the change in phase per one revolution of the machine. Therefore, the unwrapped phase has to be adjusted to one revolution. This phase is divided by the number of pulses

$$
\Phi/K \rightarrow \Phi.
\tag{4.50}
$$

This adjustment can be called the normalisation of the phase [4].

### 4.3.6    An Alternative Computing the Instantaneous Frequency

It is not always necessary to calculate the unwrapped phase. It is possible to use the alternative procedure for calculation of the instantaneous frequency of the modulated harmonic signal. The amplitude is calculated as previously described. The alternative method of calculation needs to enter the real and imaginary part of the analytic signal, that is, $x(t)$ and $y(t)$, respectively. Exactly the same calculation procedure is used for continuous signals.

The phase of the signal is defined by the formula

$$\Phi(t) = \arctan(y(t)/x(t)). \tag{4.51}$$

The amplitude of the analytic signal is calculated by the previously described formula

$$A(t) = \sqrt{x^2(t) + y^2(t)}. \tag{4.52}$$

The first derivative of the phase with respect to time results in the angular frequency

$$\omega(t) = \frac{d\Phi(t)}{dt} = \frac{d\left(\arctan(y(t)/x(t))\right)}{dt} = \frac{\dfrac{dx(t)}{dt}y(t) - x(t)\dfrac{dy(t)}{dt}}{x^2(t) + y^2(t)}. \tag{4.53}$$

The phase is calculated by integrating the angular velocity with respect to time

$$\Phi(t) = \int_0^t \omega(\tau)\,d\tau. \tag{4.54}$$

This procedure contains calculation of the first derivative of the real and imaginary part of the analytic signal with respect to time. A digital filter for numerical differentiation of the input signal is possible to design, but this filter has a time delay. The transformation of signals into the frequency domain is an easy way of avoiding the delay of the digital filter. This calculation method in the frequency domain does not cause delay of the signal as in the case of digital filters. The calculation of the first derivative is replaced by multiplying $j\omega$. and the Fourier transform of the input signal.

The demonstration example is in Figure 4.19. The frequency of the swept sine signal is increased from the initial value of 10 Hz to the final value of 30 Hz according to a linear function of time. This example also shows the effect of a time window. Using the Hanning window smoothes the graph at the beginning and end of the time interval for which the frequency is calculated. Examples in this book are typically selected such that the value of the first and last sample of the finite time signal are smoothly connected. The DFT or FFT calculations assume that the input signal is one period of the endless periodic signal. This example includes a jump at the edges of the period of the saw signal. The beginning and the end of the graph is therefore influenced by calculation so that it contains a transient part.

### 4.3.7    Envelope Analysis

The process of the amplitude demodulation is also referred to as envelope analysis. The calculation of the envelope signal can be used for signals that are stationary, nonstationary

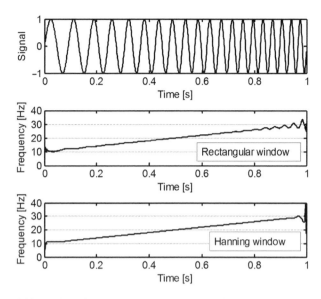

**Figure 4.19**   Calculation of the signal frequency using the alternative method.

and especially transient. The transient signal is a name for the fading or decaying signals. The envelope may facilitate the evaluation of the parameters as a relative damping or decay constant.

The envelope is mathematically the absolute value of the doubled sum of the phasors which rotate in the positive direction. Of course components with large amplitude have a big impact on the length of the resulting phasor. From the viewpoint of the frequency spectrum the envelope or the amplitude modulation signal can be calculated for a signal which is either broadband or narrowband. The envelope of the broadband signal combines the effect of all spectrum components. However, only a few frequency components, that is, the carrier component of the associate sideband components can be chosen for the amplitude demodulation. Some frequency analysers are equipped with a function for calculating the envelopes of an input signal which is divided into several frequency bands.

The envelope of the amplitude-modulated signal is equal to a factor that multiplies the carrier component.

$$E(t) = A \left( 1 + \beta_{AM} \cos \left( \omega_{AM} t + \varphi_{AM} \right) \right). \tag{4.55}$$

Another example will involve a signal that is common in the diagnostics of machines. The envelope analysis is a popular technique for diagnostics of rolling bearings. The vibration signal which is produced by faulty bearings contains the series of broadband bursts of a high frequency, which are excited by striking a local fault on the bearing inner or outer race or on the rolling elements. Each shock excites the high frequency resonances which are slightly damped but the ringing ends come before the next shock. The frequency of this ringing is not as relevant as the frequency of the repetition of the bursts. This frequency of the bursts is important for localisation of the fault of the bearing race. A synthetic example is given in this part of the book. The amplitude of the ringing signal may depend on the magnitude of

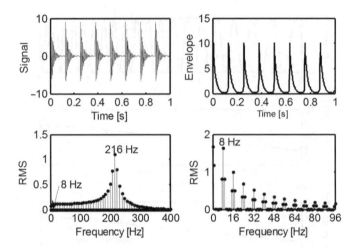

**Figure 4.20**   Calculation of the envelope of the bursts.

the impact force, which is determined by the load which transfers rolling elements [Randall]. These effects are not expected [7]. The faults which are simulated by an artificial signal assume that they concern the outer ring which is not rotating and a local fault of the bearing with the horizontal axis is in the bottom part of the ring.

It is supposed that the measured signal is composed from the repeating bursts at the ringing frequency of 216 Hz and that these bursts repeat 8 times per second. The waveform, envelope and frequency spectra are shown in Figure 4.20. The frequency spectrum of the simulated signal is shown on the left side of the bottom in Figure 4.20. The component of the resonant frequency is dominated in the frequency spectrum on the right side of the upper row in Figure 4.20. The frequency spectrum is composed of harmonic components which are spaced by the repetition frequency of the bursts. The dominating magnitude is close to the resonant frequency. The frequency spectrum of the true signals contains many families of the harmonics which are spaced variously. The component corresponding to the repetition frequency of the bursts at the beginning of the frequency scale is hard to recognise. The envelope signal and its frequency spectrum enable the frequency of the bursts to be read without a problem.

The envelope analysis can also be exploited to the advantage of signals that are not stationary in time, but in fact exponentially decaying

$$x(t) = e^{-\sigma t} \cos(\omega t + \varphi).  \tag{4.56}$$

where $\sigma$ is a decay constant [s$^{-1}$], $\omega$. is an angular frequency and $\varphi$ is an initial phase.

The envelope which is represented as a semi logarithmic plot with logarithmic vertical axis is a straight line whose slope is numerically equal to the negative value of the decay constant. Because the discrete Fourier transform is used to calculate the signal envelope at a time interval of length $T$ the phasors of this signal are derived. The expansion of the exponential function of time in the form of the trigonometric series can be done in the following way

$$e^{-\sigma t} = \sum_{k=-\infty}^{+\infty} F_k \exp\left(j\frac{2\pi}{T}kt\right) = \sum_{k=-\infty}^{+\infty} \frac{1 - e^{-\sigma T}}{\sigma T + j2\pi k} e^{j\frac{2\pi}{T}kt}, \quad t \in \langle 0, T \rangle.  \tag{4.57}$$

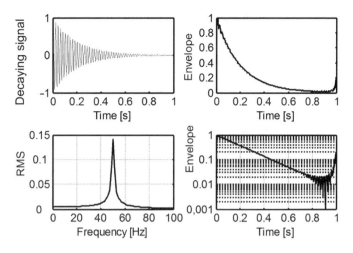

**Figure 4.21**   Calculation of the envelope of the exponentially decaying signal.

Formulas for calculating phasors will be derived using Euler's formula.

$$x^+(t) = \frac{1}{2} \sum_{k=-\infty}^{+\infty} \frac{1 - e^{-\sigma T}}{\sigma T + j2\pi k} \exp\left(+j\left(\left(\omega + \frac{2\pi}{T}k\right)t + \varphi\right)\right)$$

$$x^-(t) = \frac{1}{2} \sum_{k=-\infty}^{+\infty} \frac{1 - e^{-\sigma T}}{\sigma T + j2\pi k} \exp\left(-j\left(\left(\omega + \frac{2\pi}{T}k\right)t + \varphi\right)\right).$$

$$(4.58)$$

The envelope as the absolute value of the doubled sum of the phasors which rotate in the positive direction is as follows

$$E(t) = 2\left[\frac{1 - e^{-\sigma T}}{\sigma T} + \left(1 - e^{-\sigma T}\right) \sum_{k=1}^{+\infty} \frac{1}{(\sigma T)^2 + (2\pi k)^2} \left(\sigma T \cos\left(\frac{2\pi}{T}kt\right) - 2\pi k \sin\left(\frac{2\pi}{T}kt\right)\right)\right]$$

$$(4.59)$$

The illustrative example of the calculation of the envelope is shown in Figure 4.21. The resonance frequency is 50 Hz and the decay constant is equal to 5 s$^{-1}$. It should be noted the distortion of the envelope in approximately 20% of the time interval length at the end.

## 4.4   Cepstrum

Cepstral analysis is focused at the periodicities in the frequency spectra. It concerns a harmonics series of a base frequency or the families of sideband components of a carrier. The number of the sets of components in the frequency spectrum that are equidistantly spaced is large that they are hardly recognisable at first sight because they usually do not dominate in the spectra. This book deals with the diagnostics of the vehicle transmission units, which are based on

**Table 4.2** Cepstrum analysis.

| Original term | Derived term |
| --- | --- |
| Spectrum | Cepstrum |
| Frequency | Quefrency |
| Harmonics | Rahmonics |
| Magnitude | Gamnitude |
| Phase | Saphe |
| Filter | Lifter |
| Low pass filter | Short pass lifter |
| High pass filter | Long pass lifter |

the analysis of the frequency spectra of noise and vibration. The components of the noise or vibration spectra are caused by the defects of a rolling bearings or gear train of gearboxes. These defects do not excite the response which is a purely sinusoidal signal, but the signal with many harmonic components in the vibration spectrum that needs to be recognised in the frequency spectrum. Speech analysis also belongs to the application areas of the cepstrum analysis.

The name cepstrum was created by reversing the order of the first four letters of the English word 'spectrum'. The name of quefrency is formed in a similar fashion by changing the order of the letters in the English word 'frequency'. Also, another new word was created to describe the results of cepstrum analysis as shown in Table 4.2.

There are many definitions of the cepstrum, such as a complex cepstrum, a real cepstrum, a power cepstrum and phase cepstrum which is slightly different. The power cepstrum was defined in a 1963 paper by Bogert *et al.* [5]. We will focus on the real cepstrum, which is defined as the real part of the inverse Fourier transform of the natural logarithm of the two-sided Fourier spectrum [6]. The order of operations for calculation of the cepstrum can be written symbolically as follows

$$C(q) = \text{real} \left( \text{IFFT} \left( \log \left( |\text{FFT} \left( x(t) \right)| \right) \right) \right) \tag{4.60}$$

where $q$ is a quefrency. The direct FFT is used instead of the inverse FFT in some definitions of the cepstrum. The result of both calculations is the same apart from the scale factor. The function of the second Fourier transform is to calculate the spectrum of the spectrum. This second transform is not primarily focused on the decomposition into the sinusoidal components but the spectral lines. The calculation of the logarithm of the frequency components is intended to reduce the differences in their magnitudes. The result of the inverse Fourier transform is not a return to the time domain but to the quefrence domain since the input signal for the Fourier transform is a logarithm of the Fourier spectrum. The cepstrum is composed of components that are called rahmonics and have a similar appearance to the dominant components in the frequency spectrum of vibration or noise of rotary machines. Because these components are calculated with the use of the inverse Fourier transform their real part can be either positive or negative.

The liftering operation is similar to the filtering operation in the frequency domain where a desired quefrency range for analysis is selected by multiplying the whole cepstrum by a rectangular window at the desired position. This method is used in speech analysis.

## 4.4.1  Effect of Harmonics

The properties of the cepstrum can be analysed using the example of the spectrum, namely the expression $X_k = \log(|\mathrm{FFT}\{x_n\}|), k = 0, 1, \ldots, N-1$. The spectrum of the first example contains 16 spectral lines which are equidistantly spaced. It is the Fourier transform of an input signal of the 1s length, which is sampled at 1024 Hz, and thus contains 1024 samples. It is assumed that the input signal is periodic with a base frequency of 16 Hz, which is also the distance between adjacent components of the spectrum. According to another assumption all the components are real and have the same magnitude, that is, they represent the logarithm of the absolute value of the Fourier spectrum. The input signal for the inverse Fourier transform and the resulting cepstrum is in Figure 4.22.

The waveform of the input signal is not important for this analysis, but one cycle of the input periodical signal is repeated 16 times per second, which is exactly equal to the quefrency of 0.0625 s. The pair of periods is repeated 8 times, which is exactly equal to the quefrency of 0.125 s. This example explains the meaning of the first two rahmonics out of 16 rahmonics in the cepstrum. The components named rahmonics have the same meaning as the components called harmonics in the frequency spectrum [6].

The numerical calculation of the discrete cepstrum $c_n$ is also confirmed by the formula for the inverse Fourier transform

$$c_n = \frac{1}{N} \sum_{k=0}^{N-1} X_k \exp(j2\pi nk/N), \quad n = 0, 1, \cdots, N-1. \tag{4.61}$$

Because some values of $X_k$ are chosen equal to one and the remainder is zero, then the result of the sum of expressions $\exp(j2\pi nk/N)$ can be determined using a polygon which is

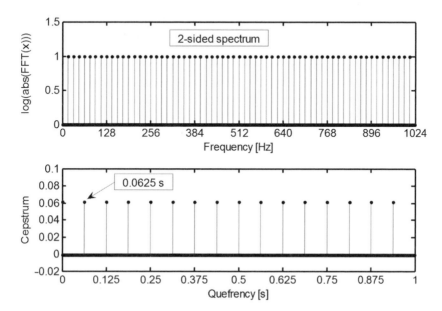

**Figure 4.22**  Cepstrum of a signal composed of only harmonic components.

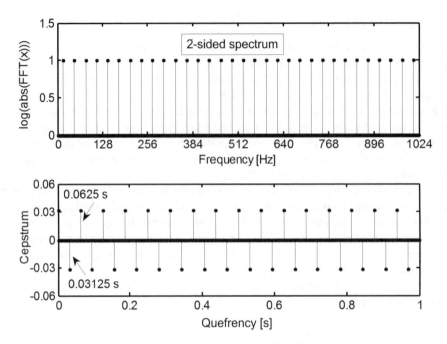

**Figure 4.23**   Cepstrum of a signal composed of only odd harmonic components.

composed of vectors as it has been demonstrated in the calculation of the discrete Fourier transform of a constant signal.

Now, attention will be drawn to the example concerning the periodic signal of the same base frequency of 16 Hz, which contains only odd harmonics. The results of the calculation in Figure 4.23 are arranged in the same way as the previous example. The string of the even harmonics of the signal has the basic frequency of 32 Hz which is twice the basic frequency of the input signal containing only the odd harmonics. In fact except for the magnitude of the nonzero components of the cepstrum, the even harmonics are missing in the spectrum. The number of positive components of the cepstra is exactly the same as in the case where the spectrum contains the even and odd harmonics together. The effect of removing the even harmonics is reflected in negative components of the cepstrum, which represent a string of the rahmonics of the basic quefrency of 0.031 25 s, that is, the frequency of 32 Hz.

Also, the inverse Fourier transform of a signal with odd harmonics can be done analytically. The choice of $X_k = 1 - 1 = 0$ for even harmonics means that the exponential terms that are multiplied by the factor $+1$ or the factor $-1$ are summed separately. The result of the sum of the exponential expressions multiplied by a factor of $-1$ gives the negative components of the discrete cepstrum.

A more realistic example is shown in Figure 4.24. It concerns calculation of the cepstrum of two signals which consists of four harmonic sinusoidal components. The frequency of these components are in the first example 50, 100, 150 and 200 Hz and in the second one 50, 150, 250 and 350 Hz. Both the signals have the same basic frequency. The amplitude of the components decreases with a frequency from highest to lowest as follows: 1, 0.5, 0.25 and 0.125. The amplitudes which are nearest in size differ by 6 dB. The ratio of the largest

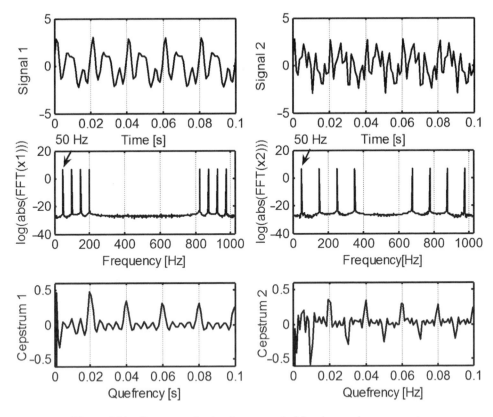

**Figure 4.24**   Cepstrum of a signal composed of four harmonic components.

amplitude to the smallest amplitude is 18 dB, which means that the smallest amplitude is suppressed almost 10 times in comparison to the largest amplitude. As is evident from the diagram in Figure 4.24, which shows the dependance of the value of log(abs(FFT(x))) on the frequency, the differences between amplitudes are considerably reduced, and therefore the assumption of identical amplitudes of harmonic and sideband components does not limit the generality of the conclusions. The five periods of the input signal are shown in Figure 4.24 as well. These examples confirm the result of analysis based on the previous idealised examples. The differences in the amplitudes of the components have no significant influence on the result of the calculation of the cepstrum.

The sampling frequency was an integer multiple of the basic frequency of the harmonic components of the idealised input periodic signal. This relationship between the sampling frequency of 1024 Hz and a frequency of 50 Hz of the basic component of the last realistic example does not apply. The effect of this phenomenon is evident in the diagrams in Figure 4.24. Yet the repetition period and the absence of the even harmonics are evident.

The nonzero components of the cepstrum for the idealised examples are either positive or negative and without an overshoot. In a realistic case, there is a small overshoot which is opposite to the main impulse when in close proximity. As will be shown below, this property is characteristic of the harmonic components of the spectrum.

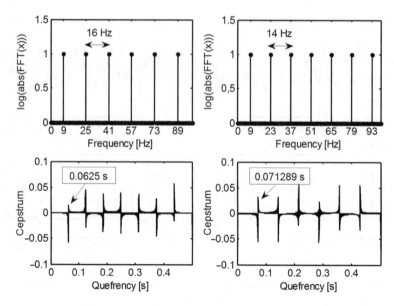

**Figure 4.25** Cepstrum of a signal composed of only sideband components.

## 4.4.2 Effect of Sideband Components

The location of the family of the sideband components in the frequency spectrum with respect to the zero frequency has more degrees of freedom. The carrier component and the distance of the side components from each are arbitrary. To study these properties idealised examples will be used. The sideband components are spaced by 16 Hz in the first example while this space of 14 Hz is assumed in the second example. The component of the frequency of 9 Hz in both the examples is the closest to the zero frequency. For simplicity, it is assumed that a set of sideband components covers the whole spectrum.

The result of the cepstrum calculation is shown in Figure 4.25. The appearance of the cepstrum components differ from the rahmonic components which correspond to the harmonic components of the frequency spectrum. The quefrency of the first dominant rahmonic component corresponds to the space between the sideband components in the frequency spectrum.

The frequency spectrum of the two sets of the sideband components is shown in Figure 4.26. The components of one set are spaced by 16 Hz while the second set are spaced by 14 Hz. The figure shows two cepstra. The cepstrum on the left side assumes the sum of the spectrum component magnitudes if they have the same frequency. In the second case, the magnitudes of all the components of the frequency spectrum are of the same value. The rahmonics of the cepstrum indicate clearly the frequency differences between harmonics, that is, 16 and 14 Hz and the corresponding quefrencies of 0.0625 s and 0.071289 s, respectively. The effect of unifying the magnitudes of the frequency spectrum give rise to rahmonics of the quefrency of 0.878 91 ms as well as the removal of even harmonics from a signal in the second example of this chapter. This quefrency corresponds to 112 Hz frequency.

The last example in the section on cepstra relates to a phase modulation signal with one carrier component of the frequency of 96 Hz. The phase modulation parameters are as follows.

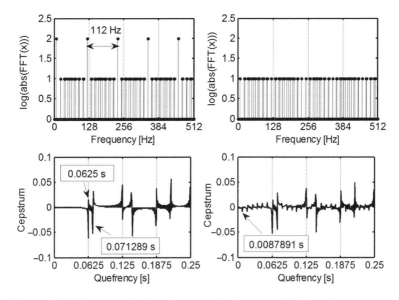

**Figure 4.26**   Cepstrum of a signal composed of two sets of sideband components.

The modulation index is equal to 0.2, and the frequency of the modulation signal is 7 Hz. The cepstrum of the modulated signal is shown in Figure 4.27. The rahmonics of the cepstrum indicate clearly the modulation frequency.

The advantage of the cepstrum is their low dependence on the characteristics of the transmission path from the source of vibrations to the location where a vibration sensor is placed. If the power spectrum of vibrations at the source is designated by $S_{xx}(f)$, then the transmission path affects the power spectrum $S_{yy}(f)$ at the location of measurement according to this formula

$$S_{yy}(f) = S_{xx}(f) \left| H_{xy}(jf) \right|^2 \tag{4.62}$$

where $H_{xy}(f)$ is the frequency response. By taking the logarithm of the last equation it is shown that the power spectrum and the transmission path of the vibration signal interact additively, that is,

$$\log S_{yy}(f) = \log S_{xx}(f) + 2 \log \left| H_{xy}(f) \right|. \tag{4.63}$$

The cepstrum of the measured signal is as follows

$$\text{IFFT}\left\{ \log S_{yy}(f) \right\} = \text{IFFT}\left\{ \log S_{xx}(f) \right\} + \text{IFFT}\left\{ 2 \log \left| H_{xy}(jf) \right| \right\}. \tag{4.64}$$

The last formula indicates an additive effect of the cepstrum of the vibration source and the inverse Fourier transform of the logarithm of the frequency response on the transmission path cepstrum of the measured signal. Because the frequency response does not usually contain the resonance peaks at the frequencies which are harmonics of a base frequency, its effect on the cepstrum is reduced.

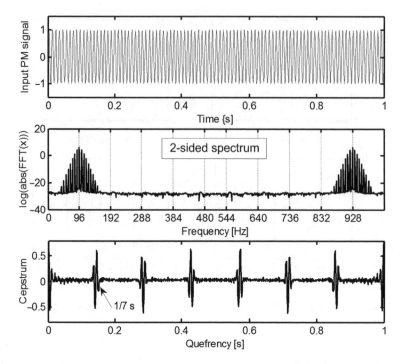

**Figure 4.27**   Cepstrum of a signal composed of a set of sideband components.

## 4.4.3   Advantages of Cepstrum

Variations of a signal over a long-period are reflected by arising components in the frequency spectra with a frequency close to zero. As the length of the time period and the frequency are mutually inversely proportional to each other, it is also quefrency inversely proportional to frequency. This property will be demonstrated on the analysis of vibration which is produced by faulty rolling bearings. A local defect on the bearing race was created with the use of discharging a charged capacitor. Measurements of the vibration signals which are generated by the bearing with a local defect and without any defects show how this defect is reflected in the spectrum and cepstrum of the vibration signal.

Tapered roller bearings on the test bench are axially loaded while the inner ring rotates at speeds close to 3000 RPM. Vibrations were measured on the outer ring in the radial direction when rotating the defect-free bearing, and then the bearing with an artificially created defect. The time histories of the acceleration signals are shown in Figure 4.28. The local imperfections excite a string of bursts that are very clear in the time record. The repetition frequency of the bursts is important for the localisation of the defects.

The frequency spectrum of both vibration signals is shown in Figure 4.29. The peak of a frequency of structural vibrations dominates at high frequency and a spectrum component of the repetition frequency of the bursts is close to zero and not recognisable due to the small amplitude. The resonance peak amplifies group harmonics or sidebands but there is no way of directly determining the distance of these spectral lines.

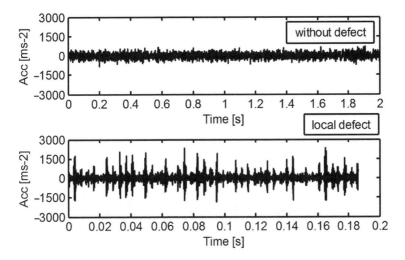

**Figure 4.28**   Vibration signals produced by healthy and faulty rolling bearings.

The response of the bearing fault records the cepstrum in Figure 4.30. The frequency for identifying the bearing defect is equal to 234 Hz. This spectrum component is located in the first 5% of the frequency range. The cepstrum of the vibration signal of the faulty bearing is clearly distinguishable from the cepstrum of the vibration signal of the healthy bearing.

Another example relates to the vibration of a gearbox. The spectrum and cepstrum of the acceleration signal is shown in Figure 4.31. The components of the spectrum are spaced by 16 Hz, which corresponds to the record of a length of 62.5 ms and the number of lines is 800, which corresponds to the record length of 2048 samples. The cepstrum components are spaced with a quefrency of 62.5 ms/2048 ≈ 0.0305 ms. The spectrum of the signal contains a set of the dominant harmonics of the fundamental frequency of 672 Hz, which are clearly recognisable. A set of the rahmonics in the cepstrum corresponds to the mentioned set of the harmonics in the frequency spectrum. The fundamental quefrency of the rahmonics is equal to

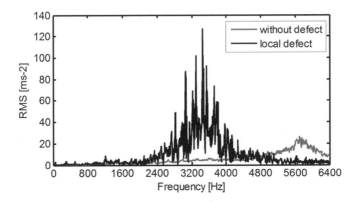

**Figure 4.29**   Vibration spectrum produced by healthy and faulty rolling bearings.

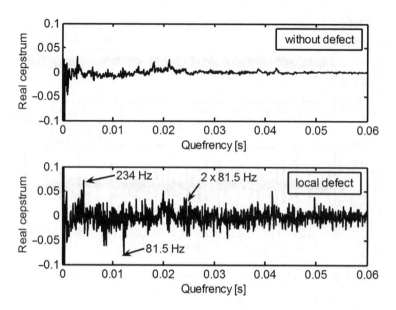

**Figure 4.30**   Cepstrum of vibration signals produced by healthy and faulty rolling bearings.

0.001 495 4 s which corresponds to the frequency of 668.7 Hz. The difference between these frequencies is less than the resolution of the frequency spectrum, which is 16 Hz.

Both tools for signal processing provide essentially the same resulting analysis of the properties of this signal.

The cepstrum is a very useful tool to detect a long periodicity of diagnostic signals which are generated in particular by the rolling bearings [7]. The cepstrum is not used for the quantitative

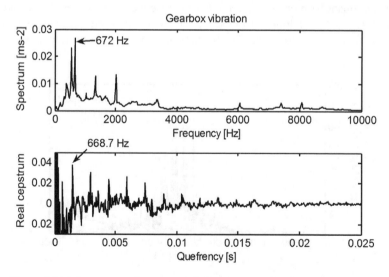

**Figure 4.31**   Spectrum and cepstrum of vibration signal produced by a gearbox.

assessment of the severity of the fault but detects the presence of faults. The origin of rahmonics in the cepstrum, that is, whether they originate from harmonics or sidebands, can be estimated according to whether they are positive or negative.

# References

[1] Tůma, J. (1993) Analysis of periodic and quasi-periodic signals in time domain, in *Proceedings of the Noise '93, St. Petersburg (Russia)*, vol. **6**, 1st edn, Auburn University, Auburn (USA), pp. 245–250.

[2] Randall, R.B. (1987) Frequency Analysis. Brüel & Kjær, Revision September 1987.

[3] Tůma, J. (2003) Phase demodulation of impulse signals in machine shaft angular vibration measurements. Proceedings of Tenth international Congress on Sound and Vibration (ICSV10), Stockholm, pp. 5005–5012.

[4] Tůma, J. (2008) Laser Doppler vibrometer and impulse signal phase demodulation in rotation uniformity measurements. Proceedings of 15th International Congress on Sound and Vibration (ICSV15), 6–10 July 2008, Daejeon, Korea.

[5] Bogert, B.P., Healy, M.J.R. and Tukey, J.W. (1963) The quefrency analysis of time series for echoes: cepstrum, pseudo autocovariance, cross-cepstrum and saphe cracking, Chapter 15, in *Proceedings of the Symposium on Time Series Analysis* (ed. M. Rosenblatt), John Wiley & Sons Inc., New York, pp. 209–243.

[6] Randall, R.B. and Hee, J. (1981) Cepstrum Analysis. Brüel & Kjaer Technical Review, No. 3.

[7] Randall, R.B. and Antoni, J. (2011) Rolling element bearing diagnostics – a tutorial. *Mechanical Systems and Signal Processing*, **25**, 485–520.

# 5

# Order Analysis

Many machines operate in the cyclic fashion, for example an IC engine with a constant number of strokes per revolution or a gear train with a constant number of mesh cycles per revolution. An analysis of noise and vibration signals produced by these machines primarily concerns the rotational speed and its harmonics. The frequency analysis of such signals is preferred in terms of the order spectra rather than the frequency spectra. As an order (abbreviation 'ord') we denote a dimensionless parameter which is a multiple of the fundamental rotational frequency. This fundamental frequency usually concerns the rotational frequency of some part of the rotating machine, such as a shaft, gear, pulley and so on. For a machine such as a gearbox we can select more than one fundamental frequency. The sampled signals are still regarded as a sequence of samples that are placed equidistantly in time. Sampling may also be equidistantly along a constant increment of rotation angle. Time is replaced by the angle of rotation or the number of rotation which is the mix of the whole number of the complete revolutions and possibly the fraction. The order spectra are evaluated using time records that are measured in dimensionless revolutions rather than seconds and the corresponding FFT spectra are measured in dimensionless orders rather than frequency in hertz. This technique is called order analysis or tracking analysis, as the rotation frequency is being tracked and used for analysis. The order tracking which is focused on only one component of the spectrum will be described in detail later in two variants, namely the quadrature mixing and Vold-Kalman filtering.

The horizontal axis in the order spectra indicates the multiple of the aforementioned fundamental rotational frequency of the rotating or reciprocating machine. If the frequency spectrum is calculated from the record, which exactly corresponds to a single complete revolution of the shaft, then the order spectrum consists only of the harmonics of the already mentioned fundamental frequency. The distance of the adjacent spectral lines of the order spectrum is equal to 1 ord. The resolution of the order spectrum of the $n$ complete successive revolutions is a fractional number equal to the $1/n$ ord. Note that the resolution of the order spectrum is the reciprocal value of the integer number of revolutions. Similarly, when determining the rotational speed by using only one pulse per the $n$ complete revolutions the sampling frequency is equal to $1/n$.

The dominating components in the frequency spectra of the noise or vibration are related to a machine rotation frequency. The rotating machines are tested at the steady-state rotation

*Vehicle Gearbox Noise and Vibration: Measurement, Signal Analysis, Signal Processing and Noise Reduction Measures*, First Edition. Jiří Tůma.
© 2014 John Wiley & Sons, Ltd. Published 2014 by John Wiley & Sons, Ltd.

or during the run-up or the coast-down operation. Location of peaks in the order spectrum in these operating conditions is the same and independent of the rotational speed. It should be mentioned that any driven unit does not rotate at a constant speed but its speed slowly varies around an average value. Rotation speed variations at the fixed signal sampling frequency cause the smearing of the dominating components in the frequency spectra. Sampling signals at the constant angle rotation increments can prevent the mentioned spectrum from smearing. The machine rotational speed measurement and identification of at least one position of a rotation shaft is of particular importance for sound and vibration analysis.

## 5.1   Speed Rotation Measurements

The rotational speed is measured in terms of the number of revolutions per minute (RPM). The simplest method for evaluation of the instantaneous rotational speed is the reciprocal value of the time interval length for a complete revolution of a machine shaft. The complete revolution is specified by a string of pulses called as a tacho-signal. The tacho-signal is generated by an optical sensor which produces a pulse which then passes a trigger attached to the surface of the shaft. The measurement range of optic sensors such as Photoelectric Tachometer Probe of the MM 0024 type is up to 20 000 RPM while laser sensors such as VLS Series Optical Speed Sensors measured up to 250 000 RPM. The monostable flip-flop of the sensor MM0024 produces a pulse of constant width therefore the measurement range is limited. The pair of adjacent pulses defines the shortest time interval for which the average rotational speed can be determined. An instant which is considered as a beginning of the shaft rotation is the time of crossing the trigger level either by the leading edge or by the falling edge of the tacho-pulse. The length of the time interval is determined by interpolation some 50 times more accurately than indicated by counting only the number of the actual sampling intervals. The accuracy is satisfying for the RPM measurement based on only the single pulse per shaft rotation. An example of a tacho-signal is shown in Figure 5.1. The procedure of identifying the beginning and end of a time interval of a single complete revolution is also shown in this figure.

The sensors which generate a string of pulses have to be connected to the DC input of the signal analyser as the mean of the signal increases when the rotational speed increases. If such a signal is connected through a high-pass filter then the portion of signal above the zero level would diminish when increasing speed, which can cause problems with setting up the trigger level when calculating RPM. The same problem with setting up the trigger level can cause the output signal of a sensor which generates pulses whose amplitude depends on the rotational speed.

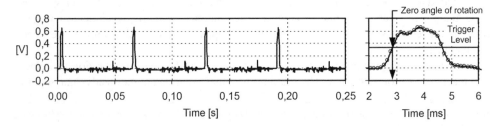

**Figure 5.1**   Example of the tacho-pulse string.

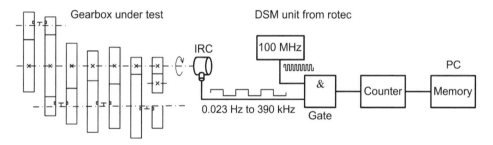

**Figure 5.2**   Rotec analyser for measurement of the instantaneous angular velocity.

The previous text has introduced the sampling frequency for measuring the speed of rotation based on the evaluation of the time interval between pulses. The sampling frequency is determined by the number of pulses per revolution. As well as the sampling frequency for the rotational speed, there is also the Nyquist frequency. According to the Nyquist-Shannon sampling theorem, we can identify the oscillation of the angular velocity at the frequency of the $m$ periods per revolution if we measure at least $2m$ pulses per revolution. Aliasing may arise also in the measurement of the rotational speed which is based on this way of sampling.

The accuracy of the estimation of the time interval length between the adjacent pulses depends on the number of samples in the time record between these pulses. If the number of time samples is small, then the error of estimation of the length of the time interval between these pulses is also large. Some smoothing may be achieved by setting the average distance between the pulses across several complete revolutions.

If it is not possible to place the trigger on the shaft whose rotation speed is to be measured, then the rotational speed is calculated by using the measured speed and gear ratio between these two shafts.

The measurement of the instantaneous angular velocity of machines is available with the use of the ROTEC analyser that supplies the Rotec GmbH Company. The measurement principle is shown in Figure 5.2. The angular velocity signal must be provided by a sensor which detects the time moments when the shaft rotates by a constant angular interval. This may be accomplished by scanning a toothed wheel with a proximity sensor or by employing an incremental rotary encoder (IRC) which is fitted to the shaft. A signal conditioning unit is required if the input signal is not of a square wave at the TTL level. The periodic times of the input pulse train are measured using an electronic oscillator of 100 MHz, 32-bit counter/timer circuit. The analyser of this type does not need an A / D converter.

The results of measurement of the time intervals are stored in a computer memory. The reciprocal value of these lengths is proportional to the instantaneous angular velocity of rotation. The cumulative sum of pulses which are generated by the high-frequency oscillator determines the angle of rotation of the shaft relative to a reference position. Angular acceleration is given by the difference of the successive values of the angular velocity.

## 5.2   Order Analysis Based on External Sampling Frequency

The calculation of DFT requires a constant number of samples per time interval between pulses that synchronises the frequency analysis with the machine rotation. In this case the

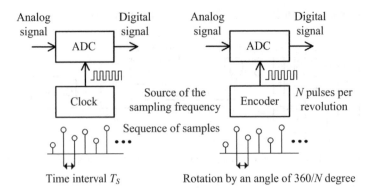

**Figure 5.3**   Methods of controlling the sample rate.

sampling frequency of the time signals such as noise and vibration must be synchronised with the rotation frequency.

The analogue signal can be sampled in two modes, see Figure 5.3. The first mode is based on a constant sampling rate. The source of the sampling frequency for the signal analysers is an oscillator which is a part of the front end. The same goes for a multifunction card. The sampling frequency is independent of the periodicities in a response of the tested object. Conversion of the samples according to the second mode is triggered by the pulses from an incremental rotary encoder (IRC). The encoder generates a string of the $N$ pulses per revolution. This mode of sampling produces a sequence of samples corresponding to the regular increment of the rotation angle. The first sampling method produces the data in the time domain while the second method produces the data in the angular domain. Synchronising the sampling rate with the frequency of rotation is ensured only by the second mode.

The number of pulses generated at the output of IRC is selected by requirements of the algorithm for calculation of DFT. For calculation of FFT this number is mostly equal to a power of two, such as 512, 1024, 2048, … In the past, the external sampling frequency was generated by the frequency multiplier (BK 5050) for multiplying the frequency of the tacho-signal by a power of two. Due to the difficulty with the phase lock loop, this method is no longer used.

The time history of signals can have a time axis on two types of scales, for example, in time measured in seconds or in angle of rotation measured in angular degrees or radians, or the number of revolutions. The number of revolutions is preferred in this book with an abbreviation 'rev'.

## 5.3   Digital Order Tracking

As already mentioned, the external trigger of sampling of the measured signal provides a perfect synchronisation between the sampling frequency and the angular speed of rotation at least at an interval corresponding to the interval between two consecutive pulses at the output of the incremental rotary encoder (IRC). It is impossible to determine fluctuations in the angular rotation speed on such data except there a time interval between two samples is measured and recorded. The encoders are often not possible to install at shaft ends. The only source of information about the rotational speed can be taken from a single pulse per revolution

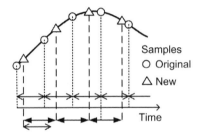

**Figure 5.4**   Interpolation of new samples.

which is obtained from an optical sensor. Because the rotational speed always varies across a small range, the numbers of samples for consecutive revolutions differ from each other at a constant sampling frequency. Synchronisation of the sampling frequency with the rotation speed is obtained by resampling the signals recorded at the constant sampling frequency. The aim of resampling is to ensure the same number of samples per revolution. For resampling the signals we use an interpolation method which determines new samples within an adjacent pair of the known samples as shown in Figure 5.4. The Lagrange or Newton interpolation polynomial can be used for interpolating new equidistantly spaced samples.

The Shannon-Nyquist theorem requires more than two samples per period due to the transient band of the anti-aliasing filter which is used to filter out the high frequency before the A/D stage of digitalisation as it was explained before. For the signal analyers of the Brüel & Kjær origin at least 2.56 samples per period are required for a sinusoidal signal . Manufacturers of the signal analysers such as the Bruel & Kjaer Company found that a sufficiently accurate resampling requires at least $2 \times 2.56 = 5.12$ samples per period of the signal components with the highest frequency of the measuring range [1]. For example, the sampling frequency of 65536 Hz enables the frequency analysis with the frequency up to 25.6 kHz while the order analysis is only with the frequency up to 12.8 kHz. Even 5.12 samples per period of the sinusoidal signal is low for sufficiently accurate interpolation, so the sampling rate is artificially doubled or quadrupled by inserting one or three samples in between adjacent samples from the A / D convertor. This operation is called oversampling. An example of four times the upsampling is shown in Figure 5.5. The original sinusoidal signal contains five samples per period. Linear interpolation would cause a considerable error. Inserting three new samples between adjacent samples decreases this interpolation error.

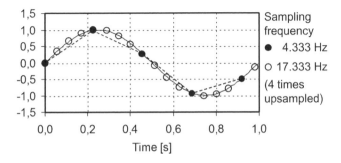

**Figure 5.5**   Four times upsampled signal.

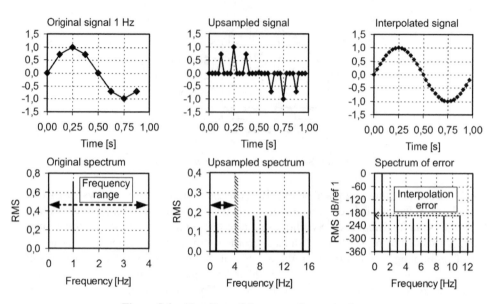

**Figure 5.6**   Algorithm of the upsampling procedure.

An algorithm for the upsampling procedure is demonstrated in Figure 5.6. The sampling frequency of the sinusoidal signal of 1 Hz will be increased four times, that is, from the original value of 8 Hz to 32 Hz by inserting three zeros between adjacent samples. Therefore, one period of the 1 Hz sinusoidal signal contains eight samples as shown in the panel on the left side in the upper row of Figure 5.6. The corresponding frequency spectrum is also on the left side, but in the bottom row. The time signal with the inserted zeros and its frequency spectrum are shown in the middle panels of both the rows. The sampling frequency of the signal in the middle and on the right side was four times greater than the original signal and is equal to 32 Hz. The two-sided frequency spectrum of the original signal contains the two conjugate components. This pair of components is repeated four times in the frequency spectrum of the signal with zeros. All the frequency spectra on the bottom row are drawn as the one-sided spectra. The frequency range of the original signal is from 0 to 4 Hz while the frequency range of the upsampled signal is from 0 to 16 Hz. All the additional components with a frequency which is greater than 4 Hz in the frequency spectrum of the signal with zeros are filtered out with the use of a digital low pass filter. The sampling frequency of the upsampled time signal is equal to 32 Hz and it is shown on the right side of the upper row. The frequency spectrum of the upsampled signal is shown in the spectrum on the right side in the bottom row. The vertical axis of this spectrum is in decibels relative to the amplitude of the original signal. The dominating component of this frequency spectrum is of 1 Hz, but the spectrum contains the background noise with RMS which is less than $-180$ dB. The value of RMS of the noise components is $10^{-9}$ times less than the amplitude of the original signal. The accuracy of the upsampling procedure approaches the accuracy of the operations with the numbers of the double precision type.

A short description of the algorithm to calculate the order spectrum is shown in the flowchart in Figure 5.7. Resampling the noise and vibration signals from machines which are operating

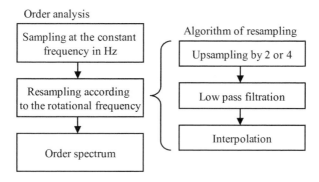

**Figure 5.7**  Flowchart for order analysis.

at the nonstationary rotational speed is also used to obtain the same number of samples per revolution, which is suitable for the synchronous filtration of the signals in the time domain.

An example of the resampling process as described above is shown in Figure 5.8. The signal is sampled at the frequency of 1024 Hz and recording takes 1 second. The signal frequency is of 150.5 Hz and amplitude is equal to unity, that is, the RMS is equal to 0.707. Frequency spectra are calculated for 400 lines, that is, the distance of the adjacent lines is of 1 Hz. The signal frequency is in the middle between the two adjacent components of the discrete frequency spectrum. No time window was used. The spectra in the top row of this figure have a linear vertical scale while the spectra in the bottom row have the logarithmic scale. The maximum of RMS in the spectrum in Hz is equal to 0.451. The frequency of the signal is a hundred times larger than the rotational speed of a virtual machine (VM). The tachosignal has a frequency of 1.505 Hz. The spectra in the left column are converted to the order spectrum. The signal is

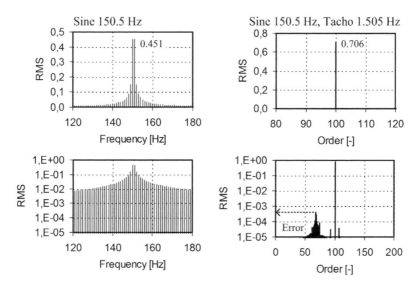

**Figure 5.8**  Sinusoidal signal 150.5 Hz, RMS 0,707, tacho 1.505 Hz.

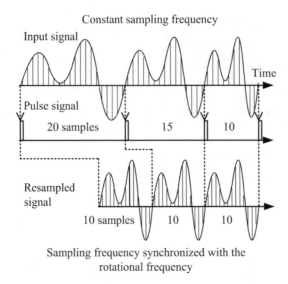

**Figure 5.9**   Resampling a signal for maintaining the same number of samples per revolution.

resampled to 512 samples for one complete revolution of VM. The maximum of RMS in the order spectrum is equal to 0.706. This value of RMS differs from the accurate value rounded to three decimal places only by 0.001. The logarithmic vertical axis of the order spectra of the resampled signal in Figure 5.8 shows that RMS of the error components is less than 0.0004 of the original signal RMS. The largest error RMS in the spectrum of the resampled signal is 65 dB less than the RMS of the original signal. It should be emphasised that the estimation of the resampling error concerned a sinusoidal signal whose frequency is constant.

Resampling accuracy depends on the width of the transition band of the digital low pass filter. The resampling procedure in the example uses the Finite Impulse Response (FIR) filter of the 128th order. Three quarters of the zeros in the calculation of the filter output reduces the number of mathematical operations for those that may not be executed at all.

The spectrum of the resampled signal for calculation of DFT contains only harmonics of the fundamental frequency, which is determined by the number of the input samples. For this reason, the most appropriate time window is rectangular. The error readout of the RMS of the spectrum components does not occur.

Order analysis is particularly important for signals that are recorded during run-up or coast down of a rotating machine, or at the rotational speed which is only slowly changed over time for example, due to changing load conditions. The individual frequency spectra are calculated from the constant length records which represent the increasing number of revolutions during the run-up of a rotary machine while the order spectra can be focused on the selected number of revolutions because the number of samples per revolution is constant as shown in Figure 5.9. Frequency components corresponding to a constant multiple of the rotational frequency are situated in the same location of the order spectrum. A special advantage of the order spectra is the fact that RMS can be readout without obvious errors in the frequency spectrum in Hz, the frequency of the sinusoidal component does not generally match with some of the discrete frequencies of the spectrum, and their position is uncertain.

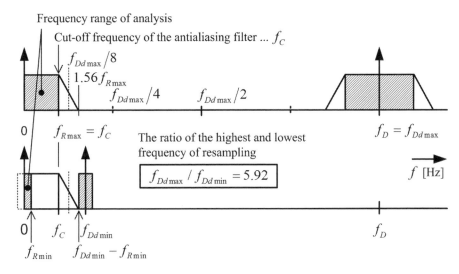

**Figure 5.10**  Permissible change of the sampling frequency without the aliasing effect.

If the rotational speed during measurements either increases or decreases it is necessary to set the sampling frequency in such a way that resampling does not cause problems with aliasing for the chosen range and the number of lines of the order spectrum. The mode of the RPM change has to be preassigned for the real-time signal analysers. If RPM increases, so the frequency of resampling also increases. It is impossible for the frequency range of the resampled signal to become larger than the passband of the anti-aliasing filter.

The following analysis concerns the frequencyof the reampled signals in hertz. Resampling with the use of interpolation is done for a signal whose sampling frequency is derived from the quadrupled sampling frequency of the A / D converter. This sampling frequency $f_D$ results from repeated halving of the original sampling frequency. The limit of values of the sampling frequency of the resampled signal with the use of interpolation, that is, the maximum $f_{Dd\,max}$ and minimum $f_{Dd\,min}$, are shown in Figure 5.10. The frequency range in the dimensionless ord and the number of lines of the order spectrum is set up for measurements and cannot be changed during this measurement after it starts.

After the stage of the repeated halving of the sampling frequency the original sampling frequency $f_D$ of the signal for interpolating new samples in between the original samples is chosen for these very reasons. Of course this frequency must conform to the highest frequency of resampling by a specified rotational speed range for the measurement. Due to the adding three zero samples the sampling frequency $f_D$ of the signal is the fourth multiple of the maximum frequency range $2.56\,f_{R\,max}$.

$$f_D = 4 \times 2.56\,f_{R\,max} = 10.24\,f_{R\,max}. \tag{5.1}$$

The maximum measurement range $f_{R\,max}$ cannot be greater then the cutoff frequency $f_C$ of the anti-aliasing filter, therefore $f_{R\,max} = f_C$. Simultaneously, the lowest frequency $f_{Dd\,min} - f_{R\,min}$ in Hz cannot be less than the maximum frequency $1.56\,f_{R\,max}$ of the transition band of the anti-aliasing filter as is shown in the lower diagram of Figure 5.10. The result is that the ratio between the maximum and minimum frequency of resampling in Hz is as follows

$$f_{Dd\,max}/f_{Dd\,min} = 5.92. \tag{5.2}$$

The appropriate choice of the above-mentioned frequency $f_D$ must conform to this ratio. If the resampling frequency varies around a steady-state value then the factor of an increase and decrease is equal to $\sqrt{5.92} = 2.43$. This measurement in the steady state mode must be set manually. Further details are described in the manuals for signal analysers supplied by the Bruel & Kjaer Company [1].

## 5.3.1 Uniformity of Rotation

When using digital order tracking the perfect synchronisation between the sampling frequency and the average angular speed of rotation over one complete revolution is all that is reached. This means that after resampling, the sampling frequency within this time interval is constant which allows detection of variations in angular velocity using the phase demodulation. For explanation of the meaning of the angular vibration it is assumed that the constant angular velocity and angle of rotation are nominal values representing the time interval of the complete revolution of a shaft. The nominal angle of rotation in degrees is a proportional part of 360 degrees which is given by the ratio of the index of the sample (initial index is zero) to the number of samples per revolution [2, 3]. The relationship between the nominal and the actual angle of rotation is explained in Figure 5.11. The calculation of the actual angle of rotation as a function of the nominal angle will be described later.

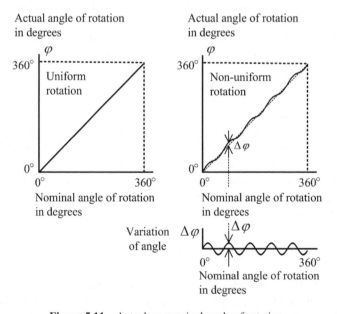

**Figure 5.11**   Actual vs. nominal angle of rotation.

So far, only the sampling was explicitly mentioned in the time or frequency domain. Digital order tracking is something between these two sampling modes which have been described above. This manner of sampling in the time interval of a complete revolution can be locally considered as sampling in the time domain because signals are sampled with a constant increment of time or nominal angle of rotation, since the time interval of one complete revolution is the shortest interval for which the average angular velocity is detectable. Globally, the speed of rotation is sampled only per a complete revolution or its integer multiple.

Digital tracking order has the disadvantage that the average speed of rotation may not be determined on the basis of the actual angle of rotation. But this method does not allow fluctuations of the average rotational speed to be determined. However, because the pulses defining the beginning and end of the complete rotation are recorded then the average rotational speed can be checked.

## 5.4   Frequency Domain Analysis Methods (Multispectral, Slice Analysis)

This subsection deals with the frequency analysis methods which result in frequency spectra in Hz or in multiples of the fundamental frequency called ord. Because this book deals mainly with noise and vibration of transmission units our interest will be focused primarily on sound pressure and vibration as acceleration, velocity or displacement. Sound pressure is measured by measuring microphones. Vibration is sensed primarily by sensors such as accelerometers, but also sensors for speed and displacement. The mode of the measurement of the frequency spectra depends on the operating conditions of the rotary machine. Some machines are rotating at a constant rotational speed, other machinery, such as vehicle drives; operate over a wide range of the rotational speed. The rotating machines are tested at a constant speed, while others, particularly a vehicle during the run-up or the coast-down.

The result of noise or vibration measurements of machines is a spectrum that is either instantaneous or averaged. The frequency of the dominating components of the spectrum depends on the machine rotational speed and therefore it is necessary to know the speed, preferably at the same time interval as the signal for the FFT calculation was recorded. The average speed is determined from the pulse signal, which starts and ends at the same time instants of the recording of sound or vibration signal to the FFT calculation. Additional information about RPM allows the spectrum components to be assigned to the source of vibration or noise. Since the sampling frequency of the signal from which the order spectrum is calculated is synchronous with the rotational speed, the extra information on speed is not necessary. The frequency of the order spectrum component in ord means the multiple of the rotational frequency is in Hz [4].

An example of vibration measurements of a gearbox is shown in Figure 5.12. The accelerometer was attached on a bearing which is close to the 27-tooth gear. This gear is rotating at 1293 RPM. These data are sufficient to calculate the gearmeshing frequency (GMF) which is equal to 528 Hz. The mating gear has 46 teeth and rotates at 759 RPM. The spectra with frequency in Hz and ord were calculated from the recorded acceleration signal. The rooth mean squares (RMS) of the spectrum components associated with meshing of the 27-tooth gear are nearly identical.

The transmission units of vehicles do not operate at constant rotational speed. Their noise or vibration tests must simulate the operational conditions at any rotational speed from a wide

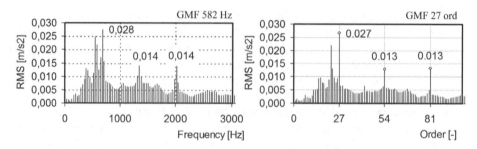

**Figure 5.12**   Frequency and order spectra of vibration of a gearbox.

working range. The tests of the truck gearbox noise are carried out in the range of the rotation speed from 1000 RPM to 2200 RPM. When testing noise of the gearboxes the rotational speed of the gearbox input shaft increases at a constant rate. The test takes about 40 seconds. During the run-up of the gearbox under the test, the sound pressure signal is recorded by measuring microphones located away from the gearbox housing at a distance of 1 m.

The multispectrum of the sound pressure level is shown in Figure 5.13. This multispectrum is composed of the instantaneous spectra recorded according to some trigger condition. The presented multispectrum is composed from the spectra differing by approximately 25 RPM. To calculate the instantaneous spectra the frequency weighting of the A type and the time window of the Hanning type was used. This multispectrum is also called the running spectrum or just a waterfall plot because of its characteristic appearance.

The multispectrum is a three-dimensional chart. This chart contains rows of peaks that are similar to a mountain range, because of their continuity in the adjacent spectra. Some rows of peaks are parallel to the RPM axis and therefore do not depend on the rotational speed and

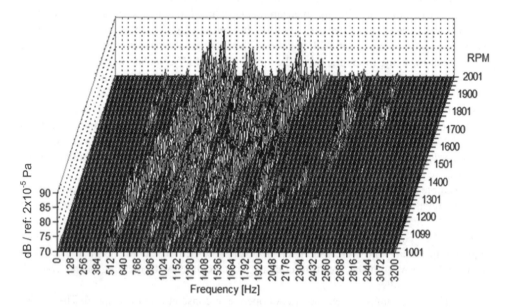

**Figure 5.13**   Waterfall multispectrum of the gearbox noise at the distance of 1 m.

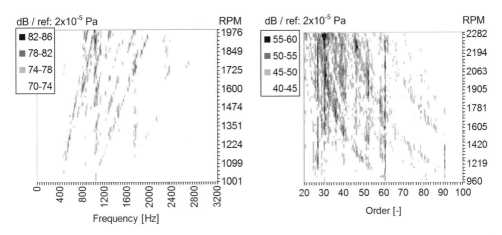

**Figure 5.14**   Contour multispectrum of the gearbox noise at the distance of 1 m.

the frequency of the peaks is constant. The origin of these peaks consists in the resonance of some parts of the gearbox structure. The other rows of peaks are diverging along radial straight lines that intersect at a virtual point corresponding to zero rotational speed. The frequency of these peaks depends proportionately on the rotational speed and therefore these vibrations are called forced. Only when the rotational speed changes can it be distinguished as resonant or forced vibrations.

The parallel and radial rows of peaks can be better recognised in the spectral map or contour plot, in which the third coordinate is replaced by colour or shade of grey. The waterfall plot in Figure 5.13 is displayed as a contour plot in the left panel of Figure 5.14. The contour plot for the order multispectrum is in the right panel of this figure. The radial rows of peaks from Figure 5.13 are transformed into the parallel rows in the order spectrum and the parallel rows of peaks from Figure 5.13 are transformed to Figure 5.14 as hyperbolas.

For testing engineers it is important to know what proportion of a specific source of vibration or noise affects the overall sound pressure level of a machine. Their attention is focused on rows of the multispectrum peaks. The dependence of the maximum RMS of these peaks on the rotational speed is called a slice. The slice can be measured directly using a tracking filter or it can be calculated from a multispectrum. The magnitude of noise or vibrations caused by the gears is equal to the magnitude of the corresponding multispectrum component of a certain frequency in ord. The multispectrum which is composed of the order spectra contains the components corresponding to the toothmeshing frequency. The frequency in ord and the number of teeth are integer numbers. The toothmeshing frequency and their harmonics in the order spectrum is equal to the integer multiples of the number of teeth if the resampling was controlled by the rotational speed of the shaft on which is mounted the gear. The magnitude of a peak in the multispectrum with frequency in Hz must be interpolated so that the equivalent value of RMS is calculated for the narrow frequency band about the toothmeshing frequency, which generally may not coincide with any of the frequencies of the frequency spectrum of this type. The frequency of a component in the order spectrum caused by the gears on the other shaft must be corrected by the gear ratio which is usually a noninteger number. The magnitude of the peak has to be interpolated in the same manner as was described. The order

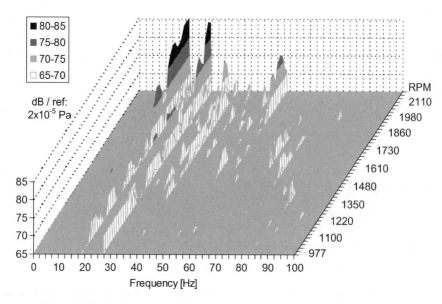

**Figure 5.15**  Order multispectrum of the gearbox noise at the distance of 1 m.

spectrum with fractional ords is best suited for interpolation. If the resampling is synchronised with a complete gear revolution then the difference between frequencies in the order spectrum is equal to 1 ord but if the resampling is synchronised with four complete revolutions then this difference is equal to 0.25 ord.

When studying the noise produced by a gear train of the gearbox the slice for certain multiples of the fundamental frequency of the machine enables the contribution of the gear train to the overall sound pressure level which is emitted by the gearbox to be assessed. An example of analysis of noise emitted by the automobile gearbox is shown in Figures 5.15 and 5.16. The kinematic diagram of the gearbox is shown in Figure 5.25. The noise of the gearbox was measured by two microphones that were placed at a distance of 1 m from the gearbox case. The test was performed in a semianechoic chamber. The torque at the input shaft was 1100 Nm. Three gears are in mesh under load. These are the gears denoted by R, hereinafter gear

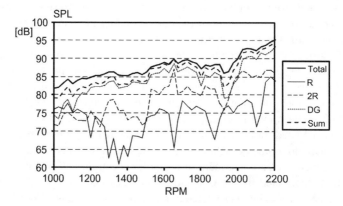

**Figure 5.16**  Slice of the multispectrum in Figure 5.15.

2R and one gear train in the drop gearbox (DR). The frequency difference between adjacent components in the order spectrum is equal to 0.25 ord. The order spectrum was calculated in relation to gearbox input shaft rotational speed and the frequency of rotation of the input shaft is formally of 1 ord. The RMS of the component of the integer ord which is equal to the number of teeth of the gear mounted on the input shaft corresponds to the magnitude of the order spectrum component and therefore it was adopted for the slice without change. Other components are not always the same as some of the frequencies of the frequency spectrum and therefore it was necessary to interpolate the corresponding RMS.

The order multispectrum of the gearbox noise is shown in Figure 5.15. The rows of peaks are parallel to the axis of RPM. The spectrum slice is also parallel. The value of the RMS is plotted against RPM in Figure 5.16, which shows the dependence of the RMS value of the spectrum components on the rotational speed of the gearbox input shaft. Each function from a set R, 2R and DG is given according to the sum of the sound powers of five harmonics and therefore after conversion in SPL this sum can be considered as the individual SPL of these gears. The sound powers of the three gear trains under load can be added together and the result is the sound power of gears in the gearbox. The sound power in $Pa^2$ is converted to SPL in dB, designed by Sum in the diagram in Figure 5.16. This sound power can be compared with the overall SPL (Total) and in this way it is possible to determine whether there is another major source of noise, such as bearings or gears in mesh freely rotating without load and without being locked.

The influence of the bandwidth on the overall SPL should be mentioned in connection with the extraction of slices which require interpolating RMS. The newly manufactured units have relatively small amplitude of the sideband components associated with the toothmeshing components and their harmonics, so the selection of the bandwidth almost does not affect the result of the interpolation of RMS. The sideband components are excited by variable load and angular vibrations which are forced from outside of the gearbox and therefore they are not related to the quality of the gears. Any wear is a matter of long-run operation. If it is necessary to select a narrow band for the interpolation of RMS then it is advisable to choose a bandwidth narrower than would correspond to the rotational speed of the gear. Smearing of the spectrum should also be respected due to changes in rotational speed during the recording of the sequence of samples for the FFT calculation. Instructions on how to choose the bandwidth to include the group of the smeared frequency components, was given in the previous chapter. The best situation is if the toothmeshing frequency in ord corresponds to a frequency of the order spectrum. In this case, the amplitude of the component is measured without error, and just uses the time window of the rectangular type [5].

The test rigs for testing the gearboxes at the end of the assembly line can be used to monitor the individual SPL for each gear train separately. It is reported that the human ear can not detect the change in a noise level less than 3 dB. Objective measurements of the noise level and sophisticated analysis allow each small response of any poor quality gears to be monitored and therefore the testing of noise can be expanded among gear manufacturers. There is no need to measure noise, just the vibrations.

## 5.5    The Use of Order Spectra for Machine Diagnostics

Order spectra are a useful tool for diagnostics of machines. The composition of the order spectra for certain types of the vibration source is shown in Figure 5.17. The order spectrum depends on the selected basic frequency. Many of the machines have several frequencies

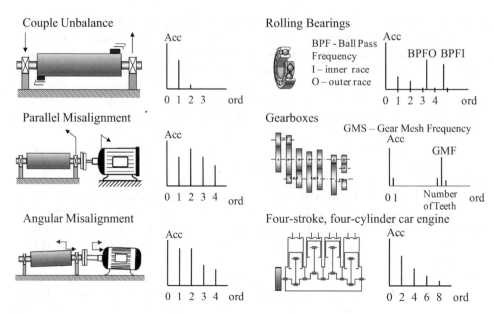

**Figure 5.17**   Order spectra in machine diagnostics.

which can be considered as fundamental. For example gearboxes contain several shafts which rotate at the different rotational speeds. For assignment of the spectrum component to the source of vibration or noise, it is important to know a value of the factor which multiplies the fundamental frequency to obtain the frequency of the spectrum component. The value of this factor also determines the possible source of vibration. With the exception of rolling bearings all the other dominating components in Figure 5.17 are harmonics of the fundamental frequency. Vibration frequency due to faults in the rolling bearings are a noninteger multiple of the rotational frequency of the inner race.

Figure 5.17 draws attention to the low multiples of the rotational frequency. Sounds of this frequency are not very audible. Noise generated by transmission units are of much higher frequency. These sources will be discussed in detail later. The order spectrum of a gearbox vibration contains a component which is denoted by GMF – Gear Mesh Frequency. GMF is calculated as the product of the number of teeth and the frequency of rotation of the gear. Engines with internal combustion, namely diesel and gasoline engines are significant sources of noise emitted by vehicles. The engine noise contains many harmonics of the firing frequency. There is only one firing per two revolutions in one cylinder of a four-stroke engine. Such four cylinder engines produce four firings per two rotations, and therefore one rotation accounts for two firings.

## 5.6   Averaging in the Time Domain

Averaging in the time domain is connected to this chapter because it makes sense to only average synchronously sampled signals. The rotation speed of machines is not stable. Stability sampling rate is many orders of magnitude higher than the stability of the rotational frequency.

The time signals which are sampled at a constant frequency need to be resampled so that each revolution corresponds to a constant number of samples. Averaging in the time domain is also referred to as synchronous filtration.

To study the excitation of vibration by acting dynamic forces, the time domain analysis based on synchronous time-domain averaging is more suitable than the frequency domain analysis. This technique is also known as a signal enhancement and has been written about by many authors (e.g. McFadden [6], Angelo [7], etc.) and can be regarded as a 'magnifying glass' (Angelo), whereby one can 'focus' on the shaft to be examined. Noise and vibration signals are sampled using a tracking technique in such a way that the length of the time record is equal to the time interval of a gear revolution. Time records are triggered synchronously by the rotation of a shaft and all the records contain the same number of samples (equal to a power of two). The time domain averaging of the samples at the same position in all the records results in a comb filtering with centre frequencies coinciding with the integer multiples of the rotational frequency [8–10].

### 5.6.1  *Principle of Averaging in the Time Domain*

Prior to averaging a signal in the time domain, the recorded signal has to be resampled as when calculating order spectra. The signal to be averaged together and the pulse signal for calculating the rotational speed are sampled at a constant sampling frequency. This is followed by digital resampling, so that each revolution contains the same number of samples. An example of measurement of the acceleration signal on the gearbox housing with the total duration of 10 seconds is shown in Figure 5.18. In parallel to the acceleration signal the string of pulses from an optical sensor that detects a trigger passing on the surface of the shaft is recorded. The rotational speed of the shaft was approximately 1500 RPM, that is, 25 Hz. Therefore the shaft rotates 248 times over 10 seconds. The rotational speed was slowly varying as is also evident in Figure 5.18. At constant sampling frequency, the number of samples per revolution is different for each revolution.

As stated above, the time records must be resampled to the same number of samples per revolution. Therefore, the horizontal axis of the resampled records cannot be scaled in time units. For presentation such resampled signals on this axis will be replaced by the number of revolutions. If this number of revolutions is equal just to unity, then the scale of the horizontal axis ranges from 0 to 1. This axis is denoted by 'Nominal revolution', because it is assumed

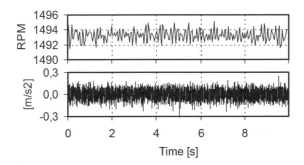

**Figure 5.18**   Acceleration signal and the varying rotational speed.

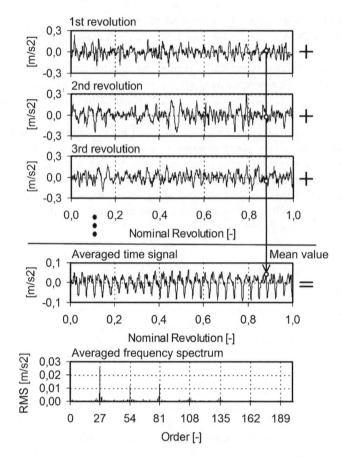

**Figure 5.19**   Principle of averaging in the time domain.

that a shaft or gear rotates at a constant angular velocity. Later it will be shown how to calculate the actual angle of rotation.

The principle of averaging in the time domain is shown in Figure 5.19. The averaged time signal contains samples whose value is equal to the mean value of all the samples having the same order in all the records corresponding to the complete revolution of the shaft. The averaged time signal is the result of averaging 248 resampled records corresponding to 248 revolutions of the 27-tooth gear.

The acceleration signal contains responses of three gear trains under load. A pinion with 27 teeth is mounted on the input shaft of the gearbox. The resampling frequency of the time signal is synchronised according to the rotational speed of the input shaft. The acceleration signals corresponding to each revolution do not allow the number of meshing cycles to be counted. The result of averaging in the time domain corresponds to the response of the 27-tooth gear. Averaging in the time domain results in not only the reduction of the RMS of an uncorrelated random signal, which is an obvious part of measured signals, but also in the suppression the response of the nonsynchronously rotating gears. The standard deviation of the noncorrelated noise after averaging is proportional to the reciprocal of the square root of the number of averages. For instance, if the number of averages equals to 100 then the RMS of random signals are reduced by 20 dB. The averaged time signal is an input for calculating the averaged frequency spectrum which contains only the dominating gear meshing components and its harmonics.

## 5.6.2 Synchronised Averaging as a Comb Filter

The synchronous time-domain averaging acts as a frequency filter. The frequency response of this filter may be analysed using the mathematical model of the filter. The process of averaging can be modelled by a moving average which is described by the following equation

$$y(t) = \frac{1}{M} \sum_{m=0}^{M-1} x(t - mT) \tag{5.3}$$

where $x(t)$ is an input signal, $y(t)$ is an output signal of the time domain filter, $T$ is a delay between samples to be used for calculating an average and $M$ is the number of the input samples which represents a number of averages. Replacing a sequence of discrete samples with a continuous function does not limit the generality of analysis. The impulse response is calculated for the Dirac delta function at the filter input

$$g(t) = \frac{1}{M} \sum_{m=0}^{M-1} \delta(t - mT). \tag{5.4}$$

The Laplace transform of the last equation transforms the time domain function into the $s$-domain function

$$G(s) = \int_0^{+\infty} g(t) \exp(-st)\, dt = \frac{1}{M} \sum_{m=0}^{M-1} \exp(-msT) = \frac{1}{M} \frac{1 - \exp(-sMT)}{1 - \exp(-sT)} \tag{5.5}$$

where $G(s)$ is the Laplace transfer function. Substituting $s$ for $j\omega$ we obtain the frequency response of the moving average filter

$$G(j\omega) = \frac{1}{M} \exp\left(-j\omega \frac{(M-1)T}{2}\right) \frac{\sin(\omega MT/2)}{\sin(\omega T/2)} \tag{5.6}$$

The magnitude of the frequency response is as follows

$$|G(j\omega)| = \frac{1}{M} \frac{|\sin(\omega MT/2)|}{|\sin(\omega T/2)|}. \tag{5.7}$$

Substituting $\omega$ for $2\pi f$ and $T$ for $1/f_0$ results in

$$|G\left(jf/f_0\right)| = \frac{1}{M}\frac{\left|\sin\left(\pi M f/f_0\right)\right|}{\left|\sin\left(\pi f/f_0\right)\right|}. \qquad (5.8)$$

where $f$ is the frequency in hertz (Hz) and $f_0$ is the synchronised frequency or the trigger frequency which is the rotational frequency in fact. The magnitude of the frequency response of the moving average filter is a periodic function, therefore

$$|G\left(jf/f_0\right)| = |G\left(j\left(f/f_0 + k\right)\right)| \qquad (5.9)$$

where $k$ is an arbitrary integer number. For $f/f_0 = 0, 1, 2, \dots$ the value of the frequency response is equal to unity. There is $M - 1$ zeros of the frequency response in between two adjacent integer multiple of the dimensionless frequency $f/f_0$

$$\sin\left(M\pi\frac{f_Z}{f_0}\right) = 0 \Rightarrow M\pi\frac{f_Z}{f_0} = k\pi \Rightarrow \frac{f_Z}{f_0} = \frac{k}{M}, \quad k = 1, 2, \dots, M - 1 \qquad (5.10)$$

An example of the dependence of the frequency response magnitude on the relative frequency $f/f_0$ is shown for $M = 10$ in Figure 5.20. Synchronous filtering does not affect the harmonics of the synchronisation frequency.

Synchronous averaging attenuates considerably the tonal components with frequencies which are a fractional multiple of the trigger frequency. The attenuation concerns the components excited by rolling bearing and by the gears rotating at the frequencies which are a fractional multiple of the trigger frequency. Except for engine timing gears, an ideal set of gears for even wear on each tooth does not have tooth numbers with a common factor other than one, which results in the mentioned fractional multiple. The gearing ratio of the timing gear train driving the camshaft is 2:1 for a 4-stroke engine and the common factor is therefore greater than one.

A part of the measured signal is an additive random noise with various distributions of power along the frequency axis. The spectrum of noise is modified by the frequency response

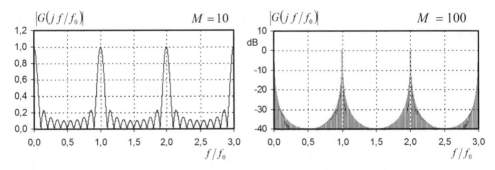

**Figure 5.20**  Magnitude of the frequency response of synchronous filtration.

of the filter of the moving average type. The power spectal density of the filter output is a product of the power spectral density at the filter input and the square of the absolute value of the filter frequency response. The positive effect of averaging in the time domain can be calculated for the additive random error of the white noise type. The variance $\eta(t)$ of the noise after averaging the white noise $\varepsilon(t)$ with the variance of $\sigma_\varepsilon^2$ at the input is given by the formula

$$\sigma_\eta^2 = E\left\{\left(\frac{1}{M}\sum_{m=0}^{M-1}\varepsilon(t-mT)\right)^2\right\} = \frac{1}{M^2}\sum_{n=0}^{M-1}\sum_{m=0}^{M-1}(E\{\varepsilon(t-mT)\varepsilon(-nT)\}). \quad (5.11)$$

The main characteristics of white noise is independent time delayed values

$$E\{\varepsilon(t-mT)\varepsilon(-nT)\} = \begin{cases} \sigma_\varepsilon^2, & m=n \\ 0, & m \neq n. \end{cases} \quad (5.12)$$

After substitution $\sigma_\varepsilon^2$ for $E\{\varepsilon(t-mT)\varepsilon(-nT)\}$ in the formula (5.11) under some condition $m=n$ it is obtained

$$\sigma_\eta^2 = \frac{1}{M^2}\sum_{m=0}^{M-1}(E\{\varepsilon(t-mT)\varepsilon(-mT)\}) = \frac{1}{M^2}M\sigma_\varepsilon^2 = \frac{\sigma_\varepsilon^2}{M}. \quad (5.13)$$

The value of the standard deviation of the random error after averaging is reduced proportionally to the reciprocal value of the square root of the number of averages.

## 5.7  Time Domain as a Tool for Gear Mesh Analysis

### 5.7.1  Averaging of Resampled Signals

The synchronous averaging in the time domain is possible only for the signals which were resampled with the use of the digital order tracking method to the same number of samples per revolution. If the record of a vibration signal concerns the response of the meshing gears, resampling of the measured signal can be synchronised with the rotational speed of each gear of the gear train. The number of the resampled signals can be as many as the gears of the gear train rotating at different speeds. The signals which are perfectly synchronised with the rotational speed of all the meshing gears enable detection of various irregularities in the gear mesh as will be shown in the next example.

The example relates to the gears with 27 and 40 teeth. As for the operating condition then the rotational speed of the pinion varies in the range of 1491 through 1495 RPM at the torque of 900 Nm. The average acceleration signal of the 27-tooth gear was presented before in Figure 5.19. The resampling of the acceleration signal was done according to the pulses generated synchronously with rotation of this gear. If it is not possible to capture directly the signal which contains the synchronisation pulses from a rotation of the second shaft then the time moments of the virtual pulses are extrapolated assuming the rotational speed of the second shaft resulting from the gear ratio. This is the procedure of how to resample and synchronously filter the vibration signal in the time domain by the rotational speed of the second gear of the gear train. The resampled signal and the result of the synchronous filtration will correspond to a complete revolution of the 40-tooth gear. As shown later, both synchronous filtration results differ.

In addition to filtering in the time domain that is synchronised by a complete revolution of the gear this operation can also be synchronised according to a repetition of meshing the same pair of the teeth. Averaging in order to reduce the influence of random errors only makes sense for the resampled signals. Resampling eliminates measurable variation in the speed of rotation. In the case of a simple gear train there are two records, the first one is synchronised according to the rotation of the pinion and the second one is synchronised according to the rotation of the wheel.

Determination of a period of repetitive contacts of the same pair of teeth on both gears in mesh will be based on the number of rotations by one circular pitch of adjacent teeth. The problem is to find the smallest positive integer as a number of the circular pitches that is divisible by both 27 and 40. The least common multiple of 27 and 40 is equal to 1080. Rotation by 1080 circular pitches is a period, when the same pair of teeth repeatedly enters in the mesh. The representation of a long period of the repetitive meshing of the same pair of teeth is preferably in a matrix form in which the rows correspond to the rotation of one or the other gear. The acceleration signal is alternating, so it is appropriate to display an immediate power, that is, the square of the acceleration sample.

An example of the measurement of vibration with the use of an accelerometer which is located on the bearing in the direction perpendicular to the gear axis regards the gear train formed by the helical gears. The results of averaging are shown in Figure 5.21 which is based on six periods of the meshing of the same teeth, 248 revolutions of the 27-tooth gear and 166 revolutions of the 40-tooth gear.

Vibration signals undoubtedly respond to minor irregularities in the surface (flank) of the teeth. The seven waves in the acceleration signal of the 40-tooth gear are clearly recognisable and can be considered to be the most interesting result of averaging due to the machining process. This response is related to the ripple surface of the teeth. The very small deviations of the tooth surface may be hardly detected by the measurement of vibration when using another method of signal processing than the averaging in the time domain. It is also evident that there is a simple relationship between cause and effect and no need to visit the frequency domain. According to the diagram on the right side in Figure 5.21 the vibration response depends on the position of the 40-tooth gear. Deviations of this gear do not depend on the 27-tooth gear in mesh. The research group around Radkowski deals with the same problem [11].

The synchronous filtration is a very convenient tool for detecting defects of the gears. The effect of the local defect on a 68-tooth gear which is in engagement with a 15-tooth gear is shown in Figure 5.22.

## 5.7.2   Effect of Load on the Averaged Acceleration Signal

It is understandable that the measured signal of acceleration on the gearbox housing depends on the operating conditions which are given by rotational speed and load. To demonstrate the effect of load on the averaged vibration signal, the response to the defect on the tooth surface is shown in Figure 5.23. The vibration signal of the loaded gears is a response to the variation of the tooth contact stiffness which overlaps the response of deviations in gear geometry or the

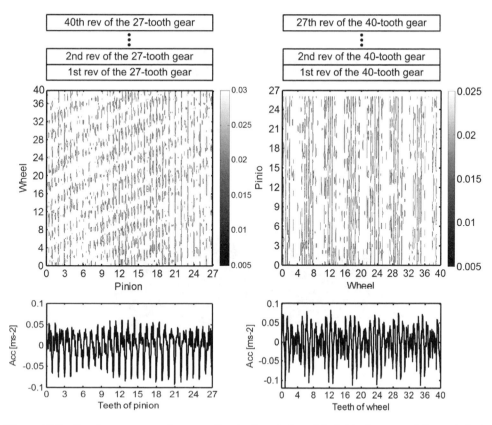

**Figure 5.21** Synchronous averaging according to the rotation of individual gears and the period of meshing the same pair of the gears.

local tooth defect. On the contrary, the vibration signal when the load is reduced is sensitive to the defects of the tooth surface.

It is well known that the vibration and noise level during the gearbox life test are maintained at a relatively stable level up to the moment when gear tooth fatigue crack starts but is a very rapidly developing process finishing with gear damage after a few minutes [12]. Continuous even wear of the teeth results in increasing the higher harmonics of the toothmeching component. More details on the diagnosis of the gear faults are given in Taylor's handbook [9].

## 5.7.3 Average Toothmesh

An example of resampling in Figure 5.19 was based on only one recorded acceleration signal and corresponding tacho pulses. The second synchronisation signal had to be replaced by the interpolated data with the use of gear ratio. In this subchapter the attention is aimed at the measurement which is made individually for each gear. This measurement with the real-time averaging was done with the use of the signal analyser which was equiped for order analysis of only one input signal and the associated tacho signal. The measurement concerns the gear train that transfers the rotational motion between the countershaft and output of the two-stage

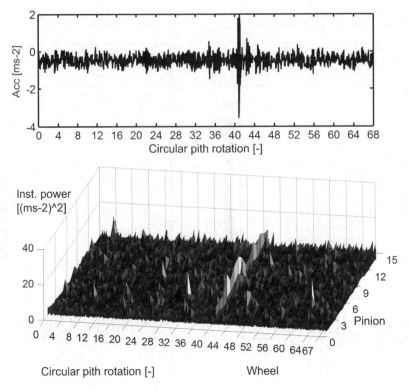

**Figure 5.22**   The response of a local defect on the 68-tooth gear.

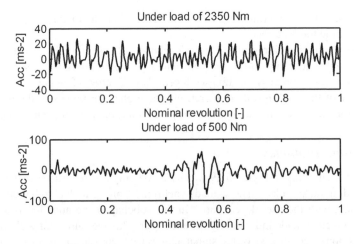

**Figure 5.23**   Gear wheel with a surface defect.

**Figure 5.24**   Back-to-back test rig at the TATRA Company a.s. and the accelerometer location.

main gearbox. This gear train is composed of the 27-tooth and 44-tooth gears. The third stage of this transmission unit of a truck is the drop gearbox. Since it was not possible to attach the trigger on either the countershaft or the output shaft the tacho pulse was measured using a trigger on the input shaft therefore the tacho signal of both the gears in mesh have to be interpolated with the use of tacho gearing.

The vibration response of the gear train and the tachosignal were recorded. The back-to-back test rig at TATRA a.s. (now TATRA TRUCK) and the location of the accelerometer is shown in Figure 5.24. The kinematic scheme of this truck gearbox is shown in Figure 5.25.

During the measurement of the acceleration signal on the gearbox housing, the input shaft rotates at the steady state speed of 1493 RPM and torque at the gearbox input shaft is 900 Nm. The 27-tooth gear rotates at the speed of $27/39 \times 1493 = 1034$ RPM. The result of acceleration

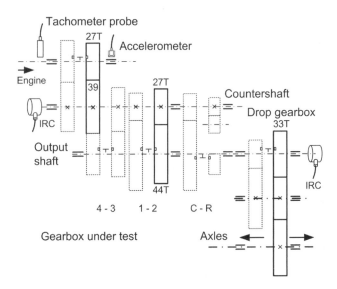

**Figure 5.25**   Kinematic scheme and sensor location for analysis of the gear train of the 27-tooth and 44-tooth gears.

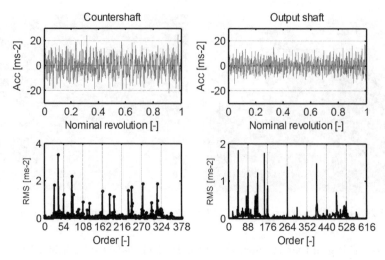

**Figure 5.26**    The result of synchronous averaging the acceleration signals.

measurements at the gearbox housing near one of the bearings and in the direction which is perpendicular to the axis of the gears is shown in Figure 5.26. The analyser was set to the frequency range of 800 ord and one record for the FFT calculation per revolution. Nominal revolution which equals to one means the complete revolution of the gear. The horizontal axis of the order spectrum contains the axis ticks at the even harmonics of the toothmeshing frequency of the gear in ord which is the same as the even multiples of the number of teeth. The order spectrum contains many harmonic components which originate from either the gear to be investigated or another gear that is also under load of torque. The signal seems to be random noise. The measured record looks better if the responses of other gears are filtered out. The signal can be divided into several parts which differ in the frequency spectrum. Filtration is performed in the frequency domain. The dependence of the weighting coefficients on the index of the spectrum components is shown in Figure 5.28. The centre frequencies of the pass bands are harmonics of the toothmeshing frequency in ord which are the integer numbers. The number of the sideband components of the toothmeshing harmonic components is reduced to 5. The result of the comb filtering is shown in Figure 5.27. This diagram shows the regularity of toothmeshing and exciting vibration by this gear.

The averaged time records corresponding to a gear rotation are corrupted by modulation signals, which are in correlation with gear geometry errors or varying gear load. The modulation signals give rise to sidebands around the carrying components with a frequency that corresponds to the harmonics of the toothmeshing frequency.The result of eliminating all the modulation effects is illustrated in Figure 5.29. Filtration was also performed in the frequency domain. Up to a scale of the horizontal axis the signals are very similar. Only the harmonic components of the toothmeshing frequency leave in the order spectrum of the signal and all the spectrum components, except harmonics of the toothmeshing frequency, are removed. The purely periodic toothmesh waveform without the modulation effects can be obtained. In view of this, the shape of this function may be illustrated by only one period corresponding to the time interval of the circular pitch rotation by one tooth.

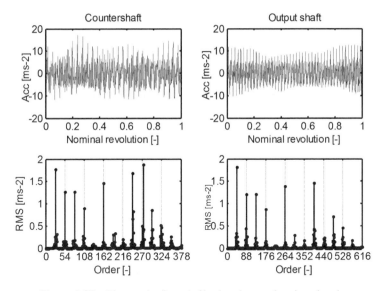

**Figure 5.27**  The result of comb filtering the acceleration signals.

In this way, filtered signals are called average toothmesh signals. The average toothmesh acceleration measured on the gearbox housing close to the shaft bearing is proportional to the dynamic forces acting between the teeth in mesh. The average toothmesh is a tool to represent the average mesh cycle [5, 13]. It can be observed that both the average toothmesh signals corresponding to meshing gears have the same shape. This fact follows from Newton's third law. The sudden increase of the acceleration signal may indicate a mechanical shock in the time moment when the teeth enters in mesh, which is a sign that the shot is not smooth and the tooth profile needs to be modified. The technique of the average toothmesh can verify this phenomenon.

The scale of the horizontal axis in nominal revolution can be changed to the number of the circular pitch rotations as is shown in Figure 5.30. The diagram contains the response of three consecutive mesh cycles to emphasise the periodicity. It is confirmed that the two different measurements contain a part of the vibration signal which is common to both the gears. The time course of both signals shows a phase shift between the two starts of these

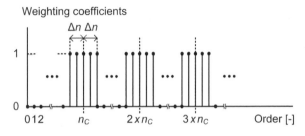

**Figure 5.28**  Pass band of the comb filter, $n$ is a centre frequency of the first pass band and $\Delta n$ is the number of the sideband components.

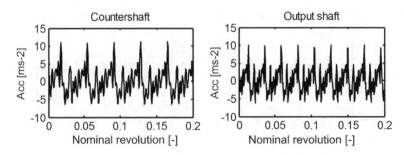

**Figure 5.29**  The result of removing the modulation effects.

two measurements with the use of order analysis instrument. However, the phase shift can be detected only in the interval of a rotation by one circular pitch. This shift can be used to correct the delay in some special measurements.

To assess a uniformity of toothmeshing during a complete gear rotation, an envelope analysis can be employed. The theory of analytical signals and the Hilbert transform are tools for envelope detecting. The envelope and the phase associated with the eight upper and lower components of the toothmeshing frequency sidebands are shown in Figure 5.31. In contrast to the similarity of the averaged toothmesh signals, the envelopes of signals generated by meshing gears may differ from each other. The meshing gears produce a signal whose harmonics of the toothmeshing frequency can be considered as a carrier whose amplitude and phase can be modulated. The phase modulation signal is also shown in Figure 5.31. The average toothmesh based on the acceleration signal is a useful tool for the analysis of the tooth contact during a tooth circular pitch rotation. The shape of the average toothmesh is almost independent on the rotational speed. The effect of RPM on the average toothmesh is shown in Figure 5.32.

## 5.7.4   Angular Vibrations

In addition to the deflection of a gear case and shaft system due to torque the teeth deflect also in the direction of the line of action. The torsional mesh stiffness of gears in mesh varies with the toothmeshing frequency in contrast to the constant and stationary stiffness of the other part of the gearbox mechanical structure. Time-varying stiffness excites so-called parametric

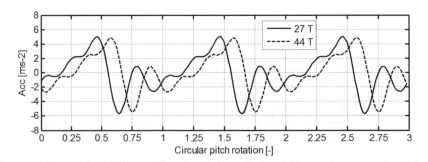

**Figure 5.30**  Three periods of the average toothmesh.

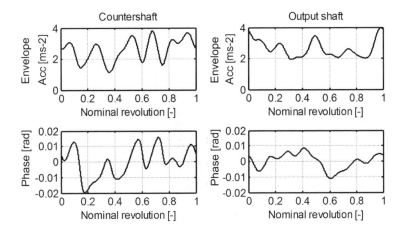

**Figure 5.31**   The result of amplitude and phase demodulation.

torsional vibrations, which can be measured. In this section the signal processing procedure will be described.

The advantage of the digital e tracking will be presented also to the analysis of the angular vibration with the use of phase demodulation. The source of the phase-modulated signal is, for example, the component of the spectrum of the frequency which is equal to the toothmeshing frequency. This measurement method has limitations of the frequency band and allows only measure angular vibrations up to frequency which is the half of the number of teeth multiple of the rotational frequency. Angular vibrations due to the toothmesh cycles can not be detected in this manner. Alternatively, this measurement can be done with the use of the phase demodulation of the pulse signal at the output of an IRC which is connected to the end of the shaft with a mounted gear.

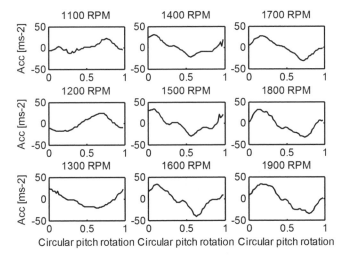

**Figure 5.32**   Effect of the RPM on the average toothmesh.

**Figure 5.33**   Mounting IRC on the end of the countershaft and output shaft of a gearbox.

The type of the IRC sensor, namely the number of pulses at the output, is selected according to the following rules. The frequency range of angular vibration measurements is sufficient up to the fifth harmonic of the toothmeshing frequency. This frequency determines the maximum bandwidth of the sidebands in the spectrum. Because we use the digital order tracking the frequency is measured in dimensionless ord which is the same number as the number of teeth. The space in the spectrum must be enough large for the upper and lower sideband of the carrying component. IRC, which generates 500 pulses, is suitable for numbers of teeth of automotive transmissions. The number of pulses per revolution determines the frequency band of the order spectrum to the frequency of 800 ord. This frequency range corresponds to 2048 samples for FFT calculation, therefore 4.096 samples per a period of the pulse signal on average. The four samples per period of pulses from IRC is sufficient.

Professional mounting rotary encoders of the type of Heidehain ERN 460-500 which are attached at the end of the countershaft and output shaft are shown in Figure 5.33. These encoders produce a string of 500 pulses and one synchronisation pulse per revolution. The encoder functionality for determining the direction of rotation is not needed. Since this measure is a continuation of measuring the average toothmesh the kinematic scheme of this truck gearbox is shown in Figure 5.25. The two encoders are ready for the measurement of the transmission error. This example concerns only the signal which is generated by IRC and connected to the countershaft.

The measured time records were resampled to 2048 samples per a complete revolution and then were averaged one hundred times. The time history of the pulse signal is shown in Figure 5.34. A series of pulses at regular amplitude, only a sampling causes the impression that the envelope is not smooth. The order spectrum of the pulse signal is shown in Figure 5.35. The carrying component of 500 ord has the upper and lower sidebands which contain two dominating families of the components. This is one of the reasons why the same order spectrum is shown twice. In the first diagram the grid ticks correspond to the 27-tooth gear and the second way of the tick location corresponds to the 39-tooth gear. Both the gears are under load and cause angular vibrations. The horizontal axis of the time history is in nominal revolution and the frequency is in dimensionless ord.

The results of phase demodulation of the IRC pulse signal and filtration by the comb bandpass filter which were described earlier are shown in Figure 5.36. Phase demodulation

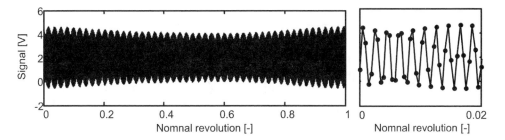

**Figure 5.34**   Time history of the pulse signal.

works with the carrier component whose frequency is equal to 500 ord and sidebands whose bandwidth is of 250 ord, therefore the total bandwidth is 500 ord about the 500-ord carrier. This frequency range can contain more than five harmonics of the toothmeshing frequency of the gears under load on the countershaft. Phase fluctuation is normalised by dividing 500, so this is a deviation from the angle of rotation which increases from 0 to 360 degrees for one complete revolution of the gear. The frequency spectrum of angular vibrations has two fundamental frequencies, which are the toothmeshing of the gears under load. The amplitude of vibrations at a frequency of 39 ord is greater than the frequency of ord 27. Separation of these two signals which differ in frequency can be done in the frequency domain. The bandpass filter is of the comb type having the pass-band centred at the frequencies which are harmonics of the toothmeshing frequency and the banwidth contains five components of the upper sideband and the same number of the lower sideband. The bottom graph on the left side in Figure 5.36 includes vibrations, which are excited by the 27-tooth gear. Angular vibrations which are excited by the 39-tooth gear are also shown in Figure 5.36. The bandwidth of the bandpass filter is 8 ord of the band width of both the sidebands. Measuring the average toothmesh and angular oscillations are part of the measurement of the transmission error, which will be described in Chapter 7 of this book.

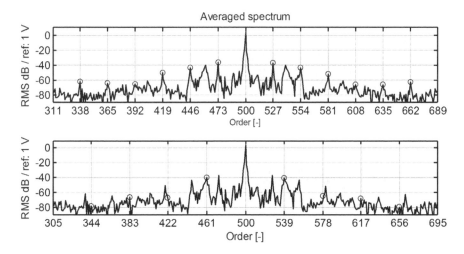

**Figure 5.35**   Order spectrum of the pulse signal.

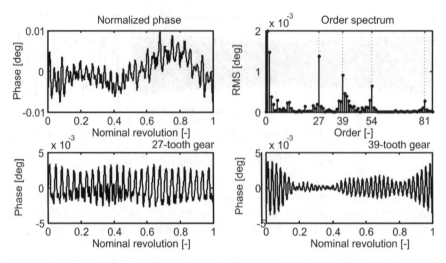

**Figure 5.36** Result of phase demodulation and bandpaas filtering centred at harmonics of 27 and 39 ord.

## 5.7.5 Effect of Averaging

Theoretically, the effect of synchronous averaging in the time domain is explained in section 5.2. The practical effect is illustrated by the measurements of vibration that have been averaged 100 times in comparison with nonaveraged signal. The first example refers to averaging the acceleration signal per revolution of the shaft with a gear. The sample is in Figure 5.37. A comparison of the instantaneous spectrum and averaged spectrum which is calculated from the time-averaged signal shows that there was a significant decrease in background noise as predicted by the mathematical model. Vibration signal is free of irregularities and is given by the sum of sinusoidal components whose origin can explain the periodicity of the machine.

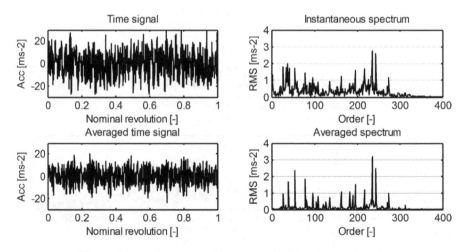

**Figure 5.37** Effect of averaging on the time signal and spectrum.

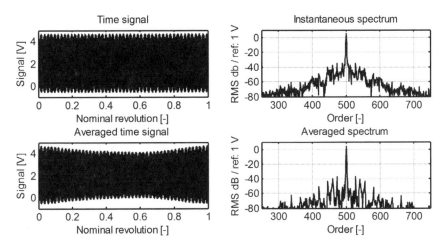

**Figure 5.38** Effect of averaging on the time signal and spectrum.

Averaging the strings of pulses has a positive effect on the result of data processing. The pulse signal itself is not a purely random signal, changing only its phase. As documented in Figure 5.38 there is a visible reduction of amplitude in the central part of the record. The real effect is, however, recognised in the spectrum of the averaged time signal. The sideband component of the spectrum contains information about variations of the phase of the pulses. Averaging increased the difference between the amplitude of the sideband components and background noise by about 20 dB which is a 10-fold difference.

The effect of averaging on the result of phase demodulation is shown in Figure 5.39. The diagrams contain angular vibrations at the toothmeshing frequency of the 27-tooth gear after averaging and without averaging. The averaged phase characterises angular vibrations better than the result which is affected by fluctuations in RPM with a very low frequency.

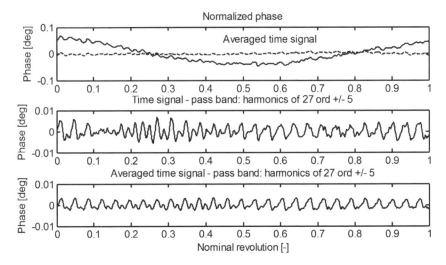

**Figure 5.39** Effect of averaging on the time signal and spectrum.

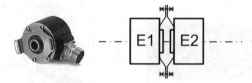

**Figure 5.40**   Arrangement of encoders to be tested.

Averaging may not relate the directly measured pulse signal to the outcome of phase demodulation. Unfortunately, this procedure is not part of the professional signal analysers yet.

## 5.7.6   Accuracy of the Incremental Rotary Encoders

The results of phase demodulation of the previous example are ranged from –0.02 to +0.02 degrees. The manufacturer of IRC, Heidenhain Company, guarantees that the accuracy of rotary encoders with line counts up to 5000, the maximum directional deviation at 20 °C ambient temperature and slow speed (scanning frequency between 1 kHz and 2 kHz), lies within ± 1/20 grating period. The encoder producing 500 pulses works with the maximum error of 0.036 degrees concerning distribution of impulses around a circle. The actual error, especially its frequency spectrum will be verified experimentally.

To compute error of the pulse distribution as a function of rotational angle a couple of encoders of the mentioned type were mounted on a shaft which ensured their identical rotational speed as shown in Figure 5.40. The difference in the angles of rotation between these encoders was assessed at the rotational speed of 1040 RPM. The pulse string generated by the encoders was sampled at the frequency of 65 536 Hz.

The measured signals and the results of calculating the difference between the phases and their order spectra are in the following figures. Changes in average speed of the encoder rotation during the measurement and changes in average speed of the encoder rotation during

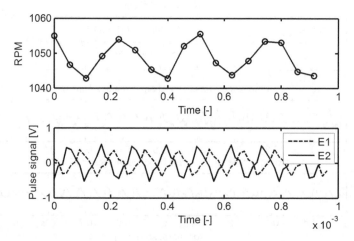

**Figure 5.41**   Average rotational speed per revolution and a piece of the time history of signals at the output of the encoders.

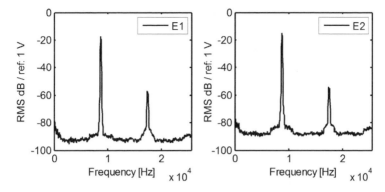

**Figure 5.42** Frequency spectrum of the signals at the output of the encoders.

the measurement and the 1-millisecond record at the outputs of the E1 and E2 encoders are shown in Figure 5.41. The frequency spectra of the two signals at the output of the encoders are shown in Figure 5.42.

The order of averaging is optional. Either the pulse signal or the result of phase demodulation can be averaged as the first. The order spectrum can be calculated from the result of the averaging in the time domain or this spectrum may be itself a result of averaging. In this case the averaging is done in the frequency domain. The algorithms *a* through c are specified by the flowchart in Figure 5.43. The effect of this choice is shown in Figure 5.44. Blocks of the flowchart and the result in Figure 5.44 connect the capital letter A through F at the right side of the some blocks.

The main result of this subchapter is an estimation of the order spectrum of the encoder errors when measuring the angle of rotation [3]. As is shown in Figure 5.45 the RMS of the encoder

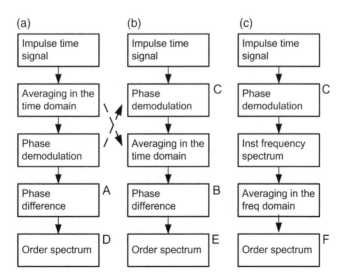

**Figure 5.43** Various arrangement of the algorithm for calculating the order spectrum.

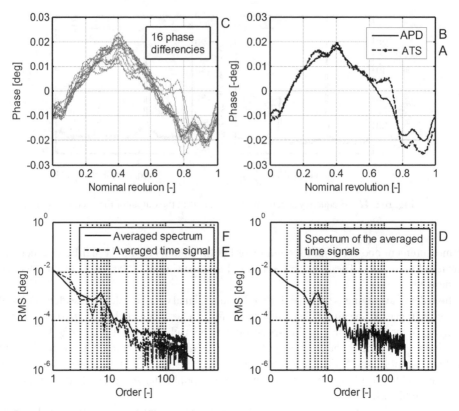

**Figure 5.44**   Effect of the order of averaging (APD – averaged phase differencies, ATS – averaged time signal).

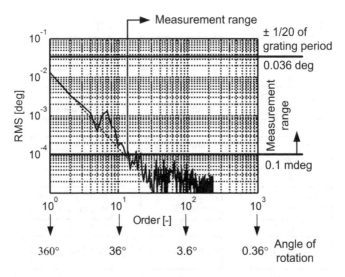

**Figure 5.45**   Measurement range of the incremental rotary encoders.

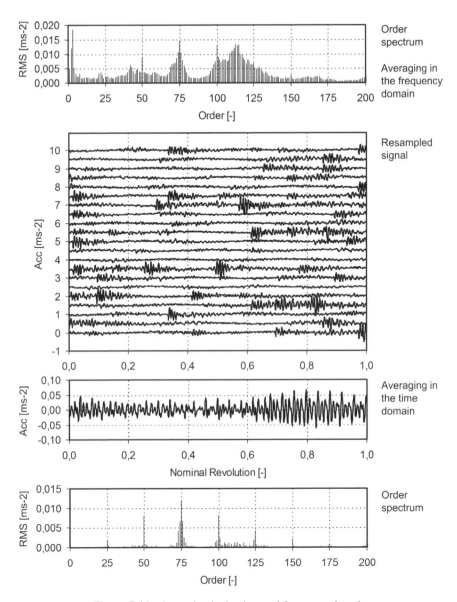

**Figure 5.46** Averaging in the time and frequency domain.

error in the order spectrum is decreased hundred times per the increase of the frequency by ten times at a low frequency up to 13 ord. The RMS of error remains approximately constant for the frequency of above 13 ord. The RMS of the error is less than 0.0001 degrees for the frequency which is greater than 13 ord. This means that it is possible to analyse angular vibrations of the gears with more than 13 teeth. Error in the distance of the pulses that are close to each other is smaller than the error of the distant pulses.

## 5.7.7 Comparison of Averaging in the Time and Frequency Domain

The effect of averaging in the time and frequency domain can be illustrated on records of the acceleration signal on the gearbox case of a lathe running at idle and producing a rattle noise. As previously described in Chapter 3, backlash and axial clearance in the teeth, idle rotation and small changes in angular speed of the driving motor causes the angular vibration with impacts of the teeth against each other. The records of the acceleration signal should be presented in a strip chart which divides the record into the intervals corresponding to one complete revolution. It is obvious that the acceleration signal must be resampled so that such drawing was possible. The result of resampling is shown in Figure 5.46.

There are two ways of calculating a representative frequency spectrum. According to the first method, the frequency spectrum is averaged in the frequency domain. According to the second method the frequency spectrum is calculated from the synchronously averaged records corresponding to complete revolutions. The differences are evident in the upper and lower graph in Figure 5.46. Because averaging was synchronised with the rotational frequency of the input shaft on which is mounted the gear with 25 teeth the harmonics of the frequency of 25 ord dominate in the frequency spectrum. Averaging in the frequency domain does not suppress the response of the other gear train in the order spectrum.

Averaging in the time domain acts as a comb filter which filters out all nonsynchronous components which have a frequency of a noninteger multiple of the rotational frequency.

Vibration signal in Figure 5.46 after time domain averaging can be repeatedly replayed as sound which corresponds to the regular operation of machines with gears. The result of replaying the original vibrations as sound gives a true broadband annoying rattle.

## References

[1]  Gade, S., Herlufsen, H., Konstantin-Hansen, H. and Wismer, N.J. (1995) Order Tracking Analysis. Brüel & Kjaer Technical Review, No. 1.
[2]  Tůma, J. (2003) Phase demodulation of impulse signals in machine shaft angular vibration measurements. Proceedings of Tenth International Congress on Sound and Vibration (ICSV10), Stockholm, pp. 5005–5012.
[3]  Tůma, J. (2008) Laser Doppler vibrometer and impulse signal phase demodulation in rotation uniformity measurements. Proceedings of 15th International Congress on Sound and Vibration (ICSV15), 6–10 July 2008, Daejeon, Korea.
[4]  Brandt, A., Lagö, T., Ahlin, K. and Tuma, J. (2005) Main principles and limitations of current order tracking methods. Proceedings of the 23rd Conference and Exposition on Structural Dynamics, IMAC-XXIII, Orlando, FL; United States; 31.1.–3.2.2005.
[5]  Tůma, J., Kuběna, R. and Nykl, V. (1994) Assessment of gear quality considering the time domain analysis of noise and vibration signals. Proceedings of 1994 Gear International Conference, Newcastle, pp. 463–468.
[6]  McFadden, P.D. (1986) Detecting fatigue cracks in gears by amplitude and phase demodulation of the meshing vibration, *ASME Journal of Vibration, Acoustics Stress and Reliability in Design*, **108**, 165–170.
[7]  Angelo, M. (1987) Vibration Monitoring of Machines. Brüel & Kjær Technical Review, No. 1.
[8]  Tůma, J. (1992) Analysis of gearbox vibration in time domain, in *Proceedings of the Euronoise '92, Imperial College London*, vol. **14**, Part 4, Book 2, 1st edn, Institute of Acoustics, London (UK), pp. 597–604. ISBN 1 873082 37 1.
[9]  Taylor, J.I. (2000) *The Gear Analysis Handbook*, Vibration Consultants, Inc., USA.
[10]  Randall, R.B. (2007) Noise and vibration data analysis, Chapter 46, in *Handbook of Noise and Vibration Control* (ed. M. Crocker), Wiley, New York, pp. 549–564.
[11]  Dybala, J., Gontarz, S. and Radkowski, S. (2007) Use of information embedded in vibroacoustic signal to crack evolution tracking of gear failure. Proceedings of the 2nd World Congress on Engineering Asset Management

(EAM) and the 4th International Conference on Condition Monitoring, 11–14 June 2007 Harrogate, UK, pp. 553–564.

[12] Begg, C.D., Byington, C.S. and Kenneth, P.M. (2000) Dynamic simulation of mechanical fault transition. Proccedings of the 54th Meeting of the Society for Machinery Failure Prevention Technology, Virginia Beach, VA, May 1–4, 2000, pp. 203–212.

[13] Tůma, J. (1993) Analysis of periodic and quasi-periodic signals in time domain, in *Proceedings of the Noise '93, St. Petersburg (Russia)*, vol. **6**, 1st edn, Auburn University, Auburn (USA), pp. 245–250.

# 6

# Tracking Filters

There is a particular class of signals which consists of harmonic components that are all (or the most dominant of them) related in frequency to the fundamental frequency, for example, the machine rotational speed [1, 2]. The frequency of this component changes proportionally to the rotational speed of the machine. These components are designated as super- or sub-harmonics (the so-called orders) of the fundamental frequency in RPM, which is measured. Test engineers focus their interest on the amplitude and phase of these orders as a function of time or more frequently as a function of RPM. This analysis technique is called the order tracking filtration. All the components of the order spectrum need not be calculated for this type of measurement.

There are two commonly used methods for this type of analysis. One method is based on the quadrature mixing and the second one uses the Vold-Kalman filter. Both methods act as a tuneable bandpass filter whose centre frequency is set up depending on the rotational speed.

The tracked harmonic spectrum component of a signal such as noise or vibration not only changes the frequency or phase but also the amplitude. The purpose of the measurement and signal processing is to determine the change in amplitude (envelope) or in phase of a selected component spectrum. This process can be described as demodulation. Since the modulation signal is contained in the sideband components to the carrying component, the main characteristics of the bandpass filter is to transfer a signal without distortion within a certain bandwidth.

## 6.1 Interpolation of the Instantaneous Rotational Speed

The main disadvantage of tracking analysis is the need to know the instantaneous rotational frequency, even for each individual sample of the signal. The accurate measurement of RPM generally only provides the average value of the rotational frequency per revolution. This contradiction can be overcome by using a spline approximation which provides a smooth function.

The input data for the interpolation of the instantaneous speed of rotation include revolutions per minute which are designated by $rpm_n$, with associated time moments $t_n$ that are assigned

*Vehicle Gearbox Noise and Vibration: Measurement, Signal Analysis, Signal Processing and Noise Reduction Measures*,
First Edition. Jiří Tůma.
© 2014 John Wiley & Sons, Ltd. Published 2014 by John Wiley & Sons, Ltd.

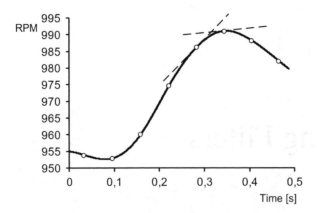

**Figure 6.1**   Cubic spline interpolation.

to this speed. The rotation speed can be measured as the number of the complete revolutions for the selected time interval. In this case the time instants are equidistant. The time interval may be a minute, it may be less than a minute. Another method consists in measuring the length of the time interval for one complete revolution as was described in Chapter 5. The spline interpolation of a data set $(t_n, rpm_n)$, $n = 0, 1, 2, \ldots, N_{rpm} - 1$ is a form of interpolation by a piecewise polynomial $P_n(t)$ called a spline between all the pairs of the predefined points $(t_{n-1}, rpm_{n-1})$ and $(t_n, rpm_n)$ for all $n$,$n = 1, 2, \ldots, N_{rpm}$. The spline interpolation is preferred over polynomial interpolation because the interpolation is continuous everywhere, also at the predefined points $(t_n, rpm_n)$, $n = 1, 2, \ldots, N_{rpm} - 1$ which are the known values of RPM. It is requested that the polynomial approximation $rpm = P_n(t)$ was true for $rpm_0 = P_1(t_0)$, $rpm_n = P_n(t_n)$ and, simultaneously, for all $n$, $1 \leq n \leq N_{rpm} - 1$ the edges of the spline segments have a common tangent

$$rpm_n = P_n(t_n), \frac{dP(t_n)}{dt} = \frac{dP_{n+1}(t_n)}{dt}, \frac{d^2P(t_n)}{dt^2} = \frac{d^2P_{n+1}(t_n)}{dt^2}. \tag{6.1}$$

This can only be achieved when using polynomials of degree 3 or higher. The classical solution is to use polynomials of degree 3, this is the case of 'Cubic splines'. An example using the cubic spline approximation is shown in Figure 6.1. A detailed description of the calculation algorithm can be found in the literature.

## 6.2   Quadrature Mixing as a Method for Amplitude and Phase Demodulation

A quadrature mixing is a demodulation method which is based on multiplication of the demodulated signal and two harmonic signals shifted against each other by $\pi/2$. The first product is considered to be a real part of a complex signal while the second one is an imaginary part. The frequency of the harmonic signal is identical to the frequency of the component whose amplitude is to be tracked. The effect of the quadrature mixing on the frequency spectrum is a shift of its frequency scale so that the frequency of the tracked component changes to

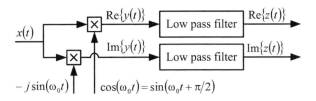

**Figure 6.2**    Quadrature mixing.

zero. The mixing is thus a multiplication of two signals, that is, a nonlinear operation, which causes frequency shifting signals into a new frequency range. The principle of this method of demodulation is described by the following calculation and it is shown in Figure 6.2.

The modulated harmonic signal $x(t)$ is considered to be in so-called envelope-and-phase form if it is written by the following formula

$$x(t) = A(t) \sin \left( \omega_0 t + \Phi(t) \right) \tag{6.2}$$

where $A(t)$ is an amplitude as a function of time, $\Phi(t)$ is a phase as well as a function of time, $\omega_0$ is a carrier frequency in radians per second and $f_0 = \omega_0/2\pi$ is a carrier frequency in hertz. The function $A(t)$ can be considered as an envelope of the component to be tracked and $\Phi(t)$ again as a phase modulation of this component. Let the maximum frequency of the amplitude and phase modulation signals be designated by $f_M$, where it is assumed that the cut-off frequency of the modulation signals is not greater than half the carrier frequency. The reason for this limitation of the modulation signal frequency will result from the possibility of separating the modulation signals from the carrier component.

The quadrature-carrier form of the signal is as follows

$$x(t) = I(t) \sin \left( \omega_0 t \right) + Q(t) \cos \left( \omega_0 t \right) \tag{6.3}$$

where $I(t)$ and $Q(t)$ are modulations of the pure carrier wave $\sin \left( \omega_0 t \right)$

$$I(t) = A(t) \sin \left( \Phi(t) \right) \quad Q(t) = A(t) \cos \left( \Phi(t) \right). \tag{6.4}$$

The component that is in phase with the original carrier $\sin \left( \omega_0 t \right)$ is referred to as an in-phase component while the out-of-phase component $\cos \left( \omega_0 t \right)$ is referred to as a quadrature component.

The quadrature mixing transforms the input signal $x(t)$ into a complex signal $y(t)$

$$\begin{aligned}
\text{Re} \left\{ y(t) \right\} &= A(t) \sin \left( \omega_0 t + \Phi(t) \right) \cos \left( \omega_0 t \right) = A(t) \left( \sin \left( \Phi(t) \right) + \sin \left( 2\omega_0 t \right) \right)/2 \\
\text{Im} \left\{ y(t) \right\} &= A(t) \sin \left( \omega_0 t + \Phi(t) \right) \sin \left( \omega_0 t \right) = A(t) \left( \cos \left( \Phi(t) \right) - \cos \left( 2\omega_0 t \right) \right)/2
\end{aligned} \tag{6.5}$$

The real part of the complex signal $y(t)$ contains a DC component $A(t) \sin \left( \Phi(t) \right)/2$ and an AC component $A(t) \sin \left( 2\omega_0 t \right)/2$ with twice the frequency of the tracked component. The imaginary part of this signal $y(t)$ also contains a DC component $A(t) \cos \left( \Phi(t) \right)/2$ and an AC component $A(t) \cos \left( 2\omega_0 t \right)/2$ with twice the frequency of the tracked component. In both cases the DC component is an AC signal whose frequency spectrum does not contain components having frequencies higher than the frequency $f_M$. The frequency of the two AC components is larger than the frequency $2f_0 - f_M$. The AC and DC parts of the spectrum can be easily separated by filtration.

The component with twice the frequency compared with the tracked component can then be filtered out by the low-pass filter. The cut-off frequency of the low-pass filter is set up to preserve a frequency band around the carrier frequency. After filtration, the complex signal is designated by $z(t)$

$$\mathrm{Re}\,\{z(t)\} = A(t)\sin(\Phi(t))/2$$
$$\mathrm{Im}\,\{z(t)\} = A(t)\cos(\Phi(t))/2. \tag{6.6}$$

The envelope of the signal after demodulation $A(t)$ or the amplitude of the tracked component is resulting from the solution of two previous equations.

$$\mathrm{mag}\,(z(t)) = |z(t)| = A(t)/2 \Rightarrow A(t) = 2\,|z(t)|\,. \tag{6.7}$$

The wrapped phase $[\Phi(t)]_{\mathrm{WRAPPED}}$ is calculated in the same way

$$\mathrm{Re}\,\{z(t)\}/\mathrm{Im}\,\{z(t)\} = \tan(\Phi(t)) \Rightarrow$$
$$[\Phi(t)]_{\mathrm{WRAPPED}} = \mathrm{arctan}\,(\mathrm{Re}\,\{z(t)\}/\mathrm{Im}\,\{z(t)\})\,. \tag{6.8}$$

Unwrapping the wrapped phase will be described in other parts of the book.

Quadraturní mixing is described for the signals with continuous time. The algorithm is also applicable for the sampled signals. Low-pass filtering can be reduced through the use of decimation appropriately sampling frequency at the output of the two filters.

An example of how to demodulate the harmonic signal which has modulated the amplitude or how to calculate the envelope of this signal is demonstrated in Figure 6.3. The two-side autospectrum of the signal is shown in the same figure. The signal was generated by the formula.

$$A(t) = 1 + 0.5\sin(\omega_M t)$$
$$\Phi(t) = 0, \tag{6.9}$$

where $\omega_M$ is an angular frequency of the harmonic modulation signal whose frequency $f_M$ is equal to 2 Hz and the amplitude modulation index is equal to 0.5. The sampling frequency is equal to 64 Hz and the carrier frequency is equal to 10 Hz.

The separate mixing of the input signal with either the sine or cosine function of time is shown in Figure 6.4. Both the frequencies are identical to the frequency of the carrier component. The effect of mixing on the two-sided autospectrum is shown at the bottom panel of this figure. The full spectrum is shifted in the left direction (toward negative frequency) while the frequency of the furthest on the right carrier component in the shifted spectrum coincides with the DC component, then the furthest on the left carrier component doubles its frequency.

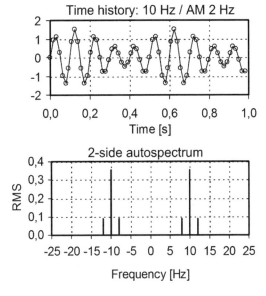

**Figure 6.3** A harmonic signal which has modulated amplitude and its two-sided autospectrum.

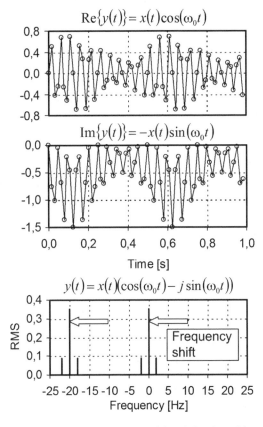

**Figure 6.4** The effect of quadrature mixing on the modulated signals and its two-sided autospectrum.

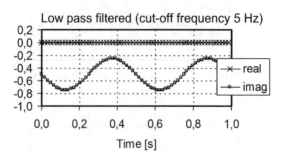

**Figure 6.5** The effect of the low-pass filter on the mixed signals.

The result of the low-pass filtration is shown in Figure 6.5. The real part is equal to zero. The absolute value of the imaginary part is the amplitude modulation signal

$$A(t) = 1 + 0.5 \sin\left(\omega_M t\right) \tag{6.10}$$

The constant frequency of the carrier component was chosen for ease of demonstration of the frequency shift. The low-pass filtration was performed using the direct and inverse Fourier transforms, so called filtration in the frequency domain. Some components of the spectrum with the absolute frequency greater than 15 Hz were set to zero. This procedure causes no phase delay. One weakness of the method is the use of the low-pass filter in the time domain. All digital filters operate with a phase delay, which depends on the filter order. The instantaneous values of the rotational speed and output of the filter can be time-shifted. This disadvantage eliminates the Vold-Kalman order tracking filter because it has the zero phase shift.

## 6.3 Kalman Filter

The Kalman filter is an algorithm that produces a statistically optimal estimate of the system state [3,4]. The input of the Kalman filter is a sequence of observations or measurements which are corrupted by errors and other inaccuracies. The role of the Kalman filter is to suppress the influence of the random error on the state estimation. The filter is named after Hungarian mathematician Rudolf E. Kálmán, who described a recursive solution in 1960. Time evolution of the state variables is described by a linear dynamic model. It is assumed that measurement errors are additive. The system state estimator is adjusted for minimum variance of the errors of the state estimation. The additional input parameters of the Kalman filter that one should be aware of prior to calculation are parameters of the state model and parameters of the probability density functions of the random errors. The model is excited by a normally distributed and uncorrelated random variable and the measurement error is also a random variable of the same type. The filter is named for Rudolf (Rudy) E. Kálmán, one of the primary developers of its theory.

The Kalman filter addresses the general problem of estimating the $P \times 1$ state vector $\mathbf{x}_n$ of a discrete time process at the $n$ step that is governed by a difference process equation (the control vector $\mathbf{u}_n$ was omitted)

$$\mathbf{x}_n = \mathbf{A}_n \mathbf{x}_{n-1} + \mathbf{v}_{1,n} \tag{6.11}$$

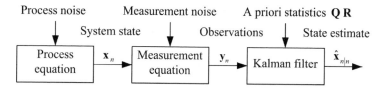

**Figure 6.6**   Estimation of the system state using the Kalman filter.

with the measurement (observation) vector $\mathbf{y}_n$ that is defined by a measurement equation, describing the observation as

$$\mathbf{y}_n = \mathbf{H}_n\mathbf{x}_n + \mathbf{v}_{2,n} \tag{6.12}$$

where random variables $\mathbf{v}_{1,n}$ and $\mathbf{v}_{2,n}$ represent the process and measurement error, respectively. It is assumed that the random variables are independent of each other and with normal probability distribution $p(\mathbf{v}_1) \sim N\left(\mathbf{0}, \mathbf{Q}_n\right)$ and $p(\mathbf{v}_2) \sim N\left(\mathbf{0}, \mathbf{R}_n\right)$. The variable $\mathbf{v}_{1,n}$ excites the process equation while the variable $\mathbf{v}_{2,n}$ excites the measurement equation. The measurement noise covariance $\mathbf{R}_n$ is usually measured or estimated prior to operation of the filter. The determination of the process noise covariance $\mathbf{Q}_n$ is generally more difficult. Sometimes a relatively simple (poor) process model can produce acceptable results if enough uncertainty is inserted into the process via the selection of $\mathbf{Q}_n$.

The matrix of the $\mathbf{A}_n$ $P \times P$ size is the state transition model which is applied to the previous state $\mathbf{x}_{n-1}$ and the $M \times P$ matrix $\mathbf{H}_n$ is the observation model which maps the true state space into the observed space.

The Kalman filter may be considered as a recursive estimator of the state vector as is shown in Figure 6.6. Although the book does not address the details of the Kalman filter, it can be noted that the filter algorithm works in two steps, namely the Time Update (called Predict) and Measurement Update (called Correct) as shown in Figure 6.7. The time update equations can also be considered as predictor equations, while the measurement update equations can be considered as corrector equations. The final estimation algorithm resembles that of a predictor-corrector algorithm for solving numerical problems.

There are two kinds of the state estimate, the a priori estimate $\hat{\mathbf{x}}_{n|n-1}$ and the a posteriori estimate $\hat{\mathbf{x}}_{n|n}$. The a priori estimate differs from the a posteriori estimate in the last measurements which were taken into account for calculation. For definitions of these conditioned

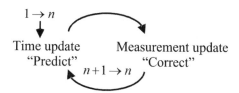

**Figure 6.7**   Kalman filter as a two-step recursive estimator.

estimates, it is necessary to introduce a set $\mathbf{Y}_n = \{\mathbf{y}_i, 1 \le i \le n\}$ of measurements. Both the estimates are defined by formulas

$$
\begin{aligned}
\text{a priori:} \quad & \hat{\mathbf{x}}_{n|n-1} = E\left\{\mathbf{x}_n | \mathbf{Y}_{n-1}\right\} \\
\text{a posteriori:} \quad & \hat{\mathbf{x}}_{n|n} = E\left\{\mathbf{x}_n | \mathbf{Y}_n\right\}
\end{aligned}
\tag{6.13}
$$

where the symbol '|' in indexes indicates conditionality of estimation. The above formulas contain conditional expected values, or mean values, which are based on conditional probability. The a priori expected value assumes that the estimate of the state vector is calculated without knowledge of the new observation while the a posteriori expected value uses the new observation for calculation. Similarly, the covariance matrices of the estimation error of the state vector are defined estimates

$$
\begin{aligned}
\text{a priori:} \quad & \mathbf{P}_{n|n-1} = E\left\{\left(\mathbf{x}_n - \hat{\mathbf{x}}_{n|n-1}\right)\left(\mathbf{x}_n(k) - \hat{\mathbf{x}}_{n|n-1}\right)^T\right\} \\
\text{a posteriori:} \quad & \mathbf{P}_{n|n} = E\left\{\left(\mathbf{x}_n - \hat{\mathbf{x}}_{n|n}\right)\left(\mathbf{x}_n - \hat{\mathbf{x}}_{n|n}\right)^T\right\}.
\end{aligned}
\tag{6.14}
$$

Taking into account the properties of the random variable $\mathbf{v}_{1,n}$ the a priori expected value $\hat{\mathbf{x}}_{n|n-1}$ is given by the recursive formula

$$
\hat{\mathbf{x}}_{n|n-1} = E\left\{\mathbf{A}_n\mathbf{x}_{n-1} + \mathbf{v}_{1,n} | \mathbf{Y}_{n-1}\right\} = \mathbf{A}_n E\left\{\mathbf{x}_{n-1} | \mathbf{Y}_{n-1}\right\} + E\left\{\mathbf{v}_{1,n} | \mathbf{Y}_{n-1}\right\} = \mathbf{A}_n\hat{\mathbf{x}}_{n-1|n-1}.
\tag{6.15}
$$

The a priori covariance matrix can be computed using the recursive formula as well

$$
\begin{aligned}
\mathbf{P}_{n|n-1} &= \text{cov}\left(\mathbf{x}_n - \hat{\mathbf{x}}_{n|n-1}\right) = \text{cov}\left(\mathbf{A}_n\mathbf{x}_{n-1} + \mathbf{v}_1(k) - \hat{\mathbf{x}}_{n|n-1}\right) \\
&= \mathbf{A}_n\text{cov}\left(\mathbf{x}_{n-1}\right)\mathbf{A}_n^T + \text{cov}\left(\mathbf{v}_{1,n}\right) = \mathbf{A}_n\mathbf{P}_{n-1|n-1}\mathbf{A}_n^T + \mathbf{Q}_n.
\end{aligned}
\tag{6.16}
$$

The a priori state estimate $\hat{\mathbf{x}}_{n|n-1}$ at step $n$ is given by knowledge of the process prior to the step $n$. After measurement of the input variable $\mathbf{y}_n$, which is done at step $n$, the a priori state estimation can be updated to the a posteriori estimate $\hat{\mathbf{x}}_{n|n}$. To design a linear estimator for calculation of the aforementioned a posteriori estimate of the state at step $n$ we introduce the difference between the actual measurement $\mathbf{y}_n$ and the estimated measurement $\mathbf{H}_n\hat{\mathbf{x}}_{n|n-1}$ which is called as an innovation of a measurement residual

$$
\tilde{\mathbf{y}}_n = \mathbf{y}_n - \mathbf{H}_n\hat{\mathbf{x}}_{n|n-1}
\tag{6.17}
$$

The new observation at step $n$ contains new information that the previous data up to step $n-1$ does not contain. After substitution we may express the measure of the new information contained in $\mathbf{y}_n$ as the measurement residual

$$
\tilde{\mathbf{y}}_n = \mathbf{H}_n\mathbf{x}_n + \mathbf{v}_{2,n} - \mathbf{H}_n\hat{\mathbf{x}}_{n|n-1} = \mathbf{H}_n\left(\mathbf{x}_n - \hat{\mathbf{x}}_{n|n-1}\right) + \mathbf{v}_{2,n} = \mathbf{H}_n\tilde{\mathbf{x}}_n^- + \mathbf{v}_{2,n}
\tag{6.18}
$$

where $\tilde{\mathbf{x}}_n^- = \mathbf{x}_n - \hat{\mathbf{x}}_{n|n-1}$ is a difference between the state vector and its prediction at step $n$ given by knowledge of the process prior to step $n$, that is, up to step $n-1$.

Without detailed justification the linear estimator of the aforementioned a posteriori estimate $\hat{\mathbf{x}}_{n|n}$ may be designed as a linear combination of the a priori estimate $\hat{\mathbf{x}}_{n|n-1}$ and the weighted innovation of the measurement residual $\tilde{\mathbf{y}}_n$.

$$\hat{\mathbf{x}}_{n|n} = \hat{\mathbf{x}}_{n|n-1} + \mathbf{K}_n \left( \mathbf{y}_n - \mathbf{H}_n \hat{\mathbf{x}}_{n|n-1} \right). \tag{6.19}$$

where $\mathbf{K}_n$ is the Kalman gain.

The a posteriori covariance matrix can be computed using the recursive formula

$$\mathbf{P}_{n|n} = \mathbf{P}_{n|n-1} - \mathbf{K}_n \mathbf{H}_n \mathbf{P}_{n|n-1} - \mathbf{P}_{n|n-1} \mathbf{H}_n^T \mathbf{K}_n^T + \mathbf{K}_n \mathbf{S}_n \mathbf{K}_n^T \tag{6.20}$$

where $\mathbf{S}_n = \text{cov} \left( \tilde{\mathbf{y}}_n \right) = \mathbf{H}_n \mathbf{P}_{n|n-1} \mathbf{H}_n^T + \mathbf{R}_n$ is a covariance of the measurement residual.

The Kalman filter is tuned by choice of gain to the minimal value of the trace $\text{tr} \left( \mathbf{P}_{n|n} \right)$ of the a posteriori error covariance matrix $\mathbf{P}_{n|n}$. Since the trace of a matrix is the sum of the elements on the main diagonal the objective function is the sum of the estimation error of the state vector elements

$$\mathbf{P}_{n|n} = \begin{bmatrix} \sigma_1^2 & & \\ & \ddots & \\ & & \sigma_P^2 \end{bmatrix} \Rightarrow \text{tr} \left( \mathbf{P}_{n|n} \right) = \sigma_1^2 + \cdots + \sigma_P^2 \tag{6.21}$$

The first partial derivative of the a posteriori error covariance matrix trace with respect to the Kalman gain has to be equal to zero

$$\frac{\partial \text{tr} \left( \mathbf{P}_{n|n} \right)}{\partial \mathbf{K}_n} = -2 \left( \mathbf{H}_n \mathbf{P}_{n|n} \right)^T + 2 \mathbf{K}_n \mathbf{S}_n = 0 \Rightarrow \mathbf{K}_n = \mathbf{P}_{n|n} \mathbf{H}_n^T \mathbf{S}_n^{-1}. \tag{6.22}$$

When substituting the definition formula of $\mathbf{S}_n$ for the covariance of the measurement residual we get the formula for calculating the optimal Kalman gain

$$\mathbf{K}_n = \mathbf{P}_{n|n-1} \mathbf{H}_n^T \left( \mathbf{H}_n \mathbf{P}_{n|n-1} \mathbf{H}_n^T + \mathbf{R}_n \right)^{-1} \tag{6.23}$$

For the optimal value of the Kalman gain the formula for calculation of the a posteriori error covariance can be simplified. Multiplying both sides of the equation for calculation of the optimal Kalman gain we get

$$\mathbf{K}_n \mathbf{S}_n = \mathbf{P}_{n|n} \mathbf{H}_n^T \Rightarrow \mathbf{K}_n \mathbf{S}_n \mathbf{K}_n^T = \mathbf{P}_{n|n} \mathbf{H}_n^T \mathbf{K}_n^T. \tag{6.24}$$

When analysing the expanded formula for the *a posteriori* error covariance

$$\mathbf{P}_{n|n} = \mathbf{P}_{n|n-1} - \mathbf{K}_n \mathbf{H}_n \mathbf{P}_{n|n-1} = \left( \mathbf{E} - \mathbf{K}_n \mathbf{H}_n \right) \mathbf{P}_{n|n-1} \tag{6.25}$$

where $\mathbf{E}$ is an identity or unit matrix. This formula is easy to use and suitable in practice. The problem with numerical stability arises due to the low precision arithmetic or a nonoptimal Kalman gain. If a nonoptimal Kalman gain is used it is not possible to use this simplification.

**Table 6.1**   Algorithm of Kalman filter.

| Phase | Formula | Description |
|---|---|---|
| Initial | $1 \rightarrow n, \hat{\mathbf{x}}_{0\vert 0} \rightarrow \hat{\mathbf{x}}_{n-1\vert n-1}, \mathbf{P}_{0\vert 0} \rightarrow \mathbf{P}_{n-1\vert n-1}$ | Initial values |
| Predict | $\hat{\mathbf{x}}_{n\vert n-1} = \mathbf{A}_n \hat{\mathbf{x}}_{n-1\vert n-1} \cdot$ | Predicted (a priori) state |
| | $\mathbf{P}_{n\vert n-1} = \mathbf{A}_n \mathbf{P}_{n-1\vert n-1} \mathbf{A}_n^T + \mathbf{Q}_n$ | Predicted (a priori) estimate covariance |
| Update | input $\rightarrow \mathbf{y}_n$ | New measurement |
| | $\tilde{\mathbf{y}}_n = \mathbf{y}_n - \mathbf{H}_n \hat{\mathbf{x}}_{n\vert n-1}$ | Innovation of the measurement residual |
| | $\mathbf{S}_n = \mathbf{H}_n \mathbf{P}_{n\vert n-1} \mathbf{H}_n^T + \mathbf{R}_n$ | Innovation (or residual) covariance |
| | $\mathbf{K}_n = \mathbf{P}_{n\vert n-1} \mathbf{H}_n^T \mathbf{S}_n^{-1}$ | Optimal Kalman gain |
| | $\hat{\mathbf{x}}_{n\vert n} = \hat{\mathbf{x}}_{n\vert n-1} + \mathbf{K}_n \tilde{\mathbf{y}}_n$ | Updated (a posteriori) state estimate |
| | $\mathbf{P}_{n\vert n} = (\mathbf{E} - \mathbf{K}_n \mathbf{H}_n) \mathbf{P}_{n\vert n-1}$ | Updated (a posteriori) estimate covariance |
| | $n + 1 \rightarrow n$ | Back to the predict phase |

As was stated before the Kalman filter has two distinct phases: prediction and updating. The algorithm is summarised in Table 6.1. The calculation is started by choosing $\hat{\mathbf{x}}_{0\vert 0}$ and $\mathbf{P}_{0\vert 0}$ and entering the input filter parameters such as $\mathbf{A}_n, \mathbf{H}_n$, $\mathbf{R}_n$ and $\mathbf{Q}_n$. The predict phase uses the state estimate from the previous step $n - 1$ to calculate an estimate of the state at the current step $n$. This predicted estimate of the state vector does not include observation from the step $n$. In the update phase, the current a priori prediction is combined with current observation information to update the state estimate.

### 6.3.1   Examples of the Use of the Kalman Filter

#### 6.3.1.1   Estimating a Random Constant

It is assumed that the true value of the unknown random constant has a standard normal probability distribution [3]. We assume that the measurement process is governed by linear difference equations

$$x_n = x_{n-1} + v_{1,n}$$
$$y_n = x_n + v_{2,n}$$

$$(6.26)$$

therefore $A = 1, H = 1$.

It is no problem to estimate the variance $R$ of the measurement noise. A small positive value of the process variance $Q$ allows us more flexibility in tuning the filter. Because we assume that the variances are not a function of $n$ the indexing is omitted. The initial value of the state variables $\hat{x}_{0\vert 0}$ may be arbitrary while the initial value of the variance $P_{0\vert 0}$ of the estimate of the initial state variables depends on the certainty of this estimate. If our estimate of $\hat{x}_{0\vert 0}$ is absolutely accurate, then $P_{0\vert 0} = 0$.

### 6.3.1.2  Estimating a Random Ramp

Testing of noise and vibrations of machines is carried out at the run up or coast down of
the rotational speed. The change in the rotational speed is proportional to the time elapsed
from the start of the test. The rate of increase or decrease of the machine rotation speed
varies randomly in a small interval. The random variation of the mentioned rate is modelled
by a centred variable $u_n$ which is assumed to be normally distributed $p(u_n) \sim N\left(0, \sigma_u^2\right)$. The
measured value of the rotational speed $y_n$ is influenced by random errors $v_{2,n}$ which is also
normally distributed $p(v_{2,n}) \sim N\left(0, \sigma_y^2\right)$ where the covariance $\sigma_{y,n}^2$ is a function of index $n$.
Due to the manner of the measurement of the rotational speed we may assume that the standard
deviation of the measurement error is inversely proportional to the time interval $\Delta t_n$ of one
complete revolution

$$\sigma_{y,n} = \sigma_{y,ref}\,\Delta t_{ref}/\Delta t_n \qquad (6.27)$$

where the index *ref* denotes the reference value of the variables.

The covariance of the measurement noise is denoted by

$$R = E\left\{v_{2,n} v_{2,n}^T\right\} = \sigma_{y,ref}^2 \left(\Delta t_{ref}/\Delta t_n\right)^2 \qquad (6.28)$$

Now, we will define the state vector for the process of increasing or decreasing the machine
speed. The first component $x_n$ of the state vector $\mathbf{x}_n$ is the average speed scaled in RPM for a
complete revolution that takes the time interval of the length $\Delta t_n$. The second component $\dot{x}_n$
is the rate of increase or decrease of the rotation speed in the change of RPM per second

$$\mathbf{x}_n = \begin{bmatrix} x_n \\ \dot{x}_n \end{bmatrix}. \qquad (6.29)$$

The process and measurement equations are as follows

$$x_n = x_{n-1} + \dot{x}_{n-1}\Delta t_n + u_n \Delta t_n$$
$$\dot{x}_n = \dot{x}_{n-1} + u_n \qquad (6.30)$$
$$y_n = x_n + v_{2,n}.$$

The matrix form of the last equations can be written as

$$\mathbf{x}_n = \mathbf{A}\mathbf{x}_{n-1} + v_{1,n}$$
$$y_n = \mathbf{H}\mathbf{x}_n + v_{2,n} \qquad (6.31)$$

where

$$\mathbf{A}_n = \begin{bmatrix} 1 & \Delta t_n \\ 0 & 1 \end{bmatrix}, \quad \mathbf{G}_n = \begin{bmatrix} \Delta t_n \\ 1 \end{bmatrix}, \quad \mathbf{H} = \begin{bmatrix} 1 & 0 \end{bmatrix}, \quad v_{1,n} = \mathbf{G}_n u_n. \qquad (6.32)$$

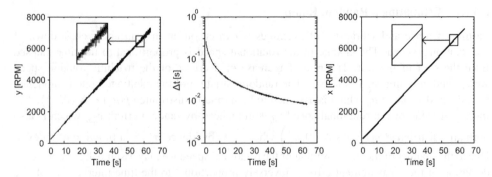

**Figure 6.8**   Filtration of a ramp signal using the Kalman filter.

It is assumed that the process noise is normally distributed $p(\mathbf{v}_{1,n}) \sim N\left(\mathbf{0}, \mathbf{Q}_n\right)$. The covariance matrix $\mathbf{Q}_n$ can be calculated using a formula

$$\mathbf{Q}_n = \mathbf{G}_n \mathbf{G}_n^T \sigma_v^2 = \begin{bmatrix} \Delta t_n^2 & \Delta t_n \\ \Delta t_n & 1 \end{bmatrix} \sigma_v^2. \tag{6.33}$$

Indexed variables emphasise their dependence on the discrete time.

An example of the filtration effect on a ramp signal is shown in Figure 6.8. The signal $y$ denotes RPM of a machine in a run up mode. At high speeds there is a reduction of the number of samples per revolution and therefore the accuracy of determination of the averaged RPM per revolution decreases. The period of the complete revolution is labelled by $\Delta t$. The signal $x$ is a result of filtering using the Kalman filter. The initial estimates are considered as a reference and its numerical values are as follows

$$\hat{\mathbf{x}}_{0|0} = \begin{bmatrix} y_1 \\ 100 \end{bmatrix}, \quad \mathbf{P}_{0|0} = \begin{bmatrix} 0.1 & 0 \\ 0 & 0.1 \end{bmatrix}. \tag{6.34}$$

## 6.4   Vold-Kalman Order Tracking Filtration

The algorithm for order tracking known as the Vold-Kalman filter was invented by Vold and Leuridan in 1993 [2, 5–7]. Kalman's name was added because the mathematical description of this filter had similarities to the Kalman filter. This new filtration technique was first incorporated into software, which was known as I-DEAS (MTS Systems Corporation) and LabShop PULSE (Brüel & Kjær Company). The first information on the calculation procedure have been described by Feldbauer and Holdrich in 2000 [8]. They introduced the procedure of how to find a solution to the equations which govern the filter behaviour and how to decouple the crossing or close orders.

The Vold-Kalman filter is a bandpass filter. The first guidebook on how to set the bandwidth of this filter, which is based on an analytical calculation, was published by the author of this book [9–11]. The Vold-Kalman filter was developed in two generations. The output of the

**Table 6.2** Comparison of terminology used to describe both filters.

| Kalman filter | Vold-Kalman filter |
| --- | --- |
| Process equation | Structural equation |
| Measurement equation | Data equation |
| State vector $\mathbf{x}_n$ | Filter single output $x_n$ |
| Measurement (observation) vector $\mathbf{y}_n$ | Filter input $y_n$, |
| Excitation vector of process equation $\mathbf{v}_{1,n}$ | Error exciting the structural equation $\varepsilon_n$ |
| Excitation vector of measurement equation $\mathbf{v}_{2,n}$ | Error which is a part of data equation $\eta_n$ |

first generation algorithm is the filtered signal while the output of the second generation is the envelope of the filtered signal of the given frequency.

As has already been mentioned the main idea of the Vold-Kalman order tracking filter (VK filter) originates from the Kalman filter. The mathematical description of the Kalman and Vold-Kalman filter seems similar but both the filters differ considerably in requirements concerning the properties of the filter output. The comparison of terminology that is used to describe the filters is stated in Table 6.2.

### 6.4.1 Data Equations of the VK-Filter

To describe the differences between the two filters we introduce vector designations for the input and output data of the Vold-Kalman filter. The filter input $x_n$ and output $y_n$ and two error terms $\eta_n$ and $\varepsilon_n$ can be arranged into vector forms as follows

$$\mathbf{y} = [y_1, \ldots, y_N]^T, \quad \mathbf{\eta} = [\eta_1, \ldots, \eta_N]^T, \quad \mathbf{x} = [x_1, \ldots, x_N]^T, \quad \mathbf{\varepsilon} = [\varepsilon_1, \ldots, \varepsilon_N]^T. \quad (6.35)$$

The Vold-Kalman filter may have multiple outputs, for example, a total of $P$ signals. In this case we can name the filter as a multiorder. The output vectors of the multiorder filter are indexed from 1 to the number of $P$

$$\mathbf{x}_k = [x_{k,1}, \ldots, x_{k,N}]^T, \quad k = 1, \ldots, P. \quad (6.36)$$

The Vold-Kalman filter is described by two equations. The first one is a data equation which is equivalent to the measurement equation of the Kalman filter. The data equation of both the generation of the VK filter decomposes the input signal $y_n$ into two additive parts: the filter output $x_n$ and an error term $\eta_n$. The output of the filter of the first generation is a narrowband signal with the centre frequency of $f_n$ ($\omega_n = 2\pi f_n$) and the bandwidth of $\Delta f_n$. The frequency $f_n$ and bandwidth $\Delta f_n$ may be a function of time (index $n$ of samples) or a constant while the output of the filter of the second generation is an envelope of the output of the filter of the first generation. The data equation for both generations of the filter has the form

$$\underset{\text{First generation}}{y_n = x_n + \eta_n}, \quad \underset{\text{Second generation}}{y_n = x_n \exp\left(j\Theta_n\right) + \eta_n}, \quad (6.37)$$

where $j$ is an imaginary unit and $\Theta_n$ is a signal phase which is defined for the sampling interval $T_S$ as follows

$$\Theta_n = \sum_{i=1}^{n} \omega_i T_S = \sum_{i=1}^{n} 2\pi f_i T_S. \qquad (6.38)$$

A complex harmonic signal $\exp\left(j\Theta_n\right)$ simplifies the solution because the result of the calculation can be focused on its real part. The effect of the multiplication of the input signal and the complex harmonic signal is that the signal component with the instantaneous frequency $f_n$ shifts toward zero frequency in the frequency spectrum. The effect of multiplication is the same as frequency shifting the original signal into the new frequency range having the zero centre frequency. The VK filter of the first generation is a type of the band-pass filter while the filter of the second-generation is a type of the low-pass filter for an envelope of the transformed signal as in the case of quadrature mixing.

The data equations for extraction of the $P$-signal components are given by the following formulas

First generation            Second generation

$$y_n = \sum_{i=1}^{P} x_{i,n} + \eta_n, \quad y_n = \sum_{k=1}^{P} x_{k,n} \exp\left(j\Theta_{k,n}\right) + \eta_n. \qquad (6.39)$$

where $\Theta_{k,n}$ is a signal phase of the $k$-th signal components to be extracted

$$\Theta_{k,n} = \sum_{i=0}^{n} \omega_{k,n} T_S \qquad (6.40)$$

The vector form of the data equation for $n = 1, \ldots, N$

First generation            Second generation

$$\begin{array}{ll} \mathbf{y} - \mathbf{x} = \boldsymbol{\eta} & P = 1 \quad \mathbf{y} - \mathbf{C}\mathbf{x} = \boldsymbol{\eta}, & P = 1 \\ \mathbf{y} - \left(\mathbf{x}_1 + \cdots + \mathbf{x}_P\right) = \boldsymbol{\eta} & P > 1 \quad \mathbf{y} - \left(\mathbf{C}_1\mathbf{x}_1 + \cdots + \mathbf{C}_P\mathbf{x}_P\right) = \boldsymbol{\eta}, \quad P > 1 \end{array} \qquad (6.41)$$

where a diagonal matrix $\mathbf{C}_k$ or $\mathbf{C}$ $(\mathbf{C} = \mathbf{C}_1)$ is defined by

$$\mathbf{C}_k = \mathrm{diag}\left\{\exp\left(j\Theta_{k,1}\right), \ldots, \exp\left(j\Theta_{k,N}\right)\right\}. \qquad (6.42)$$

To assess the difference between values of $y_n$ and $x_n$, the square of the Euclidean norm of the error vector is introduced

First generation $\quad \boldsymbol{\eta}^T\boldsymbol{\eta} = \left(\mathbf{y}^T - \mathbf{x}_1^T - \cdots - \mathbf{x}_P^T\right)\left(\mathbf{y} - \mathbf{x}_1 - \cdots - \mathbf{x}_P\right) \qquad (6.43)$

Second generation $\quad \boldsymbol{\eta}^T\boldsymbol{\eta} = \left(\mathbf{y}^T - \mathbf{x}_1^T\mathbf{C}_1^H - \cdots - \mathbf{x}_P^T\mathbf{C}_P^H\right)\left(\mathbf{y} - \mathbf{C}_1\mathbf{x}_1 - \cdots - \mathbf{C}_P\mathbf{x}_P\right). \qquad (6.44)$

where an upper index $H$ designates a Hermitian matrix $\left(\mathbf{C}_k^H = \left(\mathbf{C}_k^*\right)^T\right)$. The square $\boldsymbol{\eta}^T\boldsymbol{\eta}$ of the Euclidean norm is an $N$-multiple of the variance of the term $\eta_n$.

The meaning of the error vector $\mathbf{v}_{2,n}$ in the Kalman filter and the corresponding variable $\eta_n$ in the VK filter differ substantially. This error for the Kalman filter is an uncorrelated Gaussian noise while the error $\eta_n$ is a part of data equation in the VK filter which is simultaneously an additive component to the output of the VK filter with no prescribed properties. The error vector $\mathbf{v}_{1,n}$ in the Kalman filter and the corresponding variable $\varepsilon_n$ in the VK filter play a similar role except that the VK filter does not specify any property for the error term.

After completion of the definitions of basic quantities the VK filter can proceed to the definition of objectives filtration using the mentioned filter. The algorithm of the VK filter searches such an output $\mathbf{x} = [x_1, \ldots, x_N]^T$ of the filter in order to achieve the required ratio between the squares $\eta^T \eta$ and $\varepsilon^T \varepsilon$ of the Euclidean norm of the error vector.

Without loss of the generality, the analysis is dealing with the tracking of just one order. Under condition that the orders are not close or crossing, the multiple orders can be tracked individually in a step-by-step way.

### 6.4.2   Structural Equations of the VK-Filter First Generation

A harmonic signal with continuous time, $x(t) = A \cos(\omega t + \varphi)$, is a solution of the differential equation of the second order in which the first derivative of the time signal is missing. The harmonic oscillations are neither amplified nor damped. If the time signal is sampled at the time instants $t_n = nT_S, n = 1, 2, \ldots$, where $T_S = 1/f_S$ is a sampling interval, the harmonic oscillations are a solution of the second order difference equation with a characteristic equation having two complex conjugate roots which are equal to the following values of $z_1 = \exp(j\omega_C T_S)$ and $z_2 = \exp(-j\omega_C T_S)$, where $\omega_C$ is an angular velocity. The solution of the mentioned equation takes the form $x_n = C_1 z_1^n + C_2 z_2^n$, where $C_1, C_2$ are constants depending on the values of the first two samples. A characteristic polynomial corresponding to the characteristic equation can be written in the form $(z - z_1)(z - z_2)$ that gives the original difference equation

$$x_n - 2\cos\left(\omega_C T_S\right) x_{n-1} + x_{n-2} = 0 \tag{6.45}$$

where the coefficient of the delayed sample $x_{n-1}$ can be designated by $c_n = 2\cos\left(\omega_C T_S\right)$. The solution of the Eq. (6.45) is based on the values of the first two samples, $x_1$ and $x_2$, and the angular velocity (angular frequency) $\omega_C$. The samples beginning from the third sample can be calculated using the formula $x_n = 2\cos\left(\omega_C T_S\right) x_{n-1} - x_{n-2}$.

The structural equation for the first generation of the VK-filter takes the form

$$x_n - 2\cos\left(\omega_C T_S\right) x_{n+1} + x_{n+2} = \varepsilon_n \tag{6.46}$$

If the right side of the structural equation is equal to zero then the solution of this second order difference equation is a harmonic function. As the sine wave can slightly change its amplitude and frequency over time an unknown non-homogeneity term $\varepsilon_n$ is incorporated on the right side of the structural equation.

The system of the structural equations (6.45) containing all the samples $x_1, \ldots, x_N$ takes the following form, which can be rewritten in the matrix form

$$\begin{bmatrix} 1 & -c_1 & 1 & \ldots & 0 & 0 & 0 \\ \ldots & \ldots & \ldots & \ldots & \ldots & \ldots & \ldots \\ \ldots & \ldots & \ldots & \ldots & 1 & -c_{N-2} & 1 \end{bmatrix} \begin{bmatrix} x_1 \\ \ldots \\ x_N \end{bmatrix} = \begin{bmatrix} \varepsilon_3 \\ \ldots \\ \varepsilon_N \end{bmatrix} \tag{6.47}$$

$$\mathbf{A}\mathbf{x} = \varepsilon. \tag{6.48}$$

The number of the structural equations is equal to $N - 2$. Elements of the matrix $\mathbf{A}$ of the $(N - 2) \times N$ size depends on the values of the parameter $c_n = 2\cos(\omega_C T_S)$, which is calculated from the instantaneous angular frequency $\omega_C$. As previously mentioned, it is necessary to interpolate the average rotational speed per revolution for all the samples within the corresponding time interval.

The sum of the squares of all the unknown non-homogeneity terms $\varepsilon_n, n = 2, \ldots, N$ for the first and second-generation algorithm can be expressed as a scalar product

$$\varepsilon^T \varepsilon = \mathbf{x}\mathbf{A}^T \mathbf{A}\mathbf{x} \tag{6.49}$$

where a row vector $\varepsilon^T$ is a transpose of the column vector $\varepsilon$.

### 6.4.3 Structural Equations of the Second Generation of the VK-Filter

The complex envelope $x_n$ is the low frequency modulation of the complex carrier wave $\exp(j\Theta_n)$, where $\Theta_n$ is a phase. The low frequency modulation causes envelope smoothness. In other words envelope is locally approximated by a low order polynomial and filtered in some specific way. This condition can be expressed by a structural equation with the nonhomogeneity term $\varepsilon_n$ on the right side of this equation. The structural equation becomes an equation describing a digital filter for the input signal $\varepsilon_n$ with the output $x_n$. The filter order designates the number of the filter poles. The equations for 1-, 2-, 3- and 4-pole filter are given by

| Number of poles | Structural equation | |
|---|---|---|
| 1 | $\nabla x_n = x_n - x_{n+1} = \varepsilon_n$ | (6.50) |
| 2 | $\nabla^2 x_n = x_n - 2x_{n+1} + x_{n+2} = \varepsilon_n$ | (6.51) |
| 3 | $\nabla^3 x_n = x_n - 3x_{n+1} + 3x_{n+2} - x_{n+3} = \varepsilon_n$ | (6.52) |
| 4 | $\nabla^4 x_n = x_n - 4x_{n+1} + 6x_{n+2} - 4x_{n+3} + x_{n+4} = \varepsilon_n$ | (6.53) |

Note that the difference operator of a given order annihilates all polynomials of less than one order. The 1-pole filter corresponds to the difference equation $x_n - x_{n+1} = 0$. The root of the corresponding characteristic equation is $z_1 = 1$ and the general solution of the difference

**Table 6.3**   Matrices A of the $(N - p) \times N$ size.

| 1-pole filter, matrix size $(N - 1) \times N$: | 2-pole filter, matrix size $(N - 2) \times N$: |
|---|---|

$$\mathbf{A} = \begin{pmatrix} 1 & -1 & & & \\ & 1 & -1 & & \\ & & \cdots & \cdots & \\ & & & 1 & -1 \end{pmatrix}$$

$$\mathbf{A} = \begin{pmatrix} 1 & -2 & 1 & & \\ & 1 & -2 & 1 & \\ & & \cdots & \cdots & \cdots \\ & & & 1 & -2 & 1 \end{pmatrix}$$

| 3-pole filter, matrix size $(N - 3) \times N$: | 4-pole filter, matrix size: $(N - 4) \times N$: |
|---|---|

$$\mathbf{A} = \begin{pmatrix} 1 & -3 & 3 & -1 & & \\ & 1 & -3 & 3 & -1 & \\ & & 1 & -3 & 3 & -1 \\ & & \cdots & \cdots & \cdots & \cdots \\ & & & 1 & -3 & 3 & -1 \end{pmatrix}$$

$$\mathbf{A} = \begin{pmatrix} 1 & -4 & 6 & -4 & 1 & \\ & 1 & -4 & 6 & -4 & 1 \\ & & 1 & -4 & 6 & -4 & 1 \\ & & \cdots & \cdots & \cdots & \cdots & \cdots \\ & & & 1 & -4 & 6 & -4 & 1 \end{pmatrix}$$

equation $x_n = C z_1^n$ which is a constant. The envelope forms a piecewise constant approximation. The two pole filter relates to $x_n - 2x_{n+1} + x_{n+2} = 0$. The characteristic equation of this difference equation has a double root $z_{1,2} = 1$ and the general solution of the difference equation $x_n = a z_1^n + b n z_1^n$ which corresponds to a piecewise strait line approximation.

The system of Eqs. (6.50) to (6.53) for all the samples $x_1, \ldots, x_N$ has the same form of structural equations $\mathbf{A}\mathbf{x} = \varepsilon$ as the matrix and vector equation (6.48). The only difference is the size of matrix $\mathbf{A}$ which depends on the number $p$ of poles as is shown in Table 6.3. The matrix $\mathbf{A}$ for the 2-pole filter is of the same structure as the matrix $\mathbf{A}$ for the first generation VK-filter with the substitution 2 for $c_2$ which corresponds to $\cos(\omega_C T_S) = 1$ and therefore $\omega_C = 0$.

The sum of the squares of all the unknown nonhomogeneity terms $\varepsilon_n, n = p + 1, \ldots, N$ is given by the equation of the same form as in Eq. (6.49).

### 6.4.4   Global Solution of the Single Order Tracking Filtration

In this chapter we will deal with the problem of monitoring only one order, that is, $P = 1$. Later, we notice the problem of how to decouple the crossing or close orders, that is, $P > 1$. The vector $\mathbf{x}$ is not needed to be indexed, a subscript may be omitted.

The input data for filtering are composed of a sampled signal $y_n, n = 1, \ldots, N$ and the corresponding instantaneous frequency $f_n, n = 1, \ldots, N$ of the tracked component which is the output signal $x_n, n = 1, \ldots, N$ of the VK filter. The length of the output signal in the samples is the same length as the input signal. The input frequency corresponds to the centre frequency of the bandpass filter. The system of the data equations from (6.50) to (6.53) and the structural equations (6.48) is an underdetermined system for the unknown waveform $x_n, n = 1, \ldots, N$, and also the terms $\eta_n, n = 1, \ldots, N$ and $\varepsilon_n, n = p + 1, \ldots, N$, which are not required for the user but these unknowns must be taken into account for solution. The number of equations is equal to $2N - p$ while the total number of unknowns is equal to $3N - p$. The additional condition for the equation solution is that the variances of the nonhomogeneity terms $\varepsilon_n$ and the other sinusoidal components and background random noise $\eta_n$ have to be minimal while maintaining the given relationship between them. The global solution can be found using the standard least square technique.

The weighted sum of the particular sums (6.49) and (6.43) or (6.44) gives the loss function

$$J = r^2 \varepsilon^T \varepsilon + \eta^T \eta \tag{6.54}$$

where $r$ is a weighting factor [9, 11]. The choice of a large value for the weighting factor $r$ leads to the highly selective filtration in the frequency domain that takes a long time to converge in amplitude. In contrast, fast convergence with low frequency resolution is achieved by choosing a low value of weightin factor $r$.

The first derivative of the loss function (6.54) with respect to the vector $\mathbf{x}$ gives a condition for the minimum of this function, which is called a normal equation.

First generation VK-filter          Second generation VK-filter

$$\frac{\partial J}{\partial \mathbf{x}} = 2r^2 \mathbf{A}^T \mathbf{A} \mathbf{x} + 2(\mathbf{x} - \mathbf{y}) = 0 \quad \frac{\partial J}{\partial \mathbf{x}} = 2r^2 \mathbf{A}^T \mathbf{A} \mathbf{x} + 2\left(\mathbf{x} - \mathbf{C}^H \mathbf{y}\right) = 0 \tag{6.55}$$

The matrix equations (6.55) are of the same form except that the vector $\mathbf{y}$ is multiplied by the matrix $\mathbf{C}^H$, which shifts the frequency of the tracked components toward zero. The pass-band filter becomes the low-pass filter.

The unknown waveform in the case of the first generation VK-filter and the unknown envelope in the case of the second-generation VK-filter result from the equations (6.55)

First generation VK-filter      Second generation VK-filter

$$\mathbf{x} = \left(r^2 \mathbf{A}^T \mathbf{A} + \mathbf{E}\right)^{-1} \mathbf{y}, \quad \mathbf{x} = \left(r^2 \mathbf{A}^T \mathbf{A} + \mathbf{E}\right)^{-1} \mathbf{C}^H \mathbf{y} \tag{6.56}$$

The product of the matrixes, $\mathbf{A}^T$ and $\mathbf{A}$, gives a symmetric matrix. The matrix $\mathbf{B} = r^2 \mathbf{A}^T \mathbf{A} + \mathbf{E}$ becomes the symmetric positive definite matrix by adding the identity matrix $\mathbf{E}$ to the main diagonal. The matrix $\mathbf{B}$ consists of the limit number of the nonzero diagonals. Therefore, it is easy to invert it. The number of the nonzero diagonals of the matrix $\mathbf{B}$ for the VK-filter of the second-generation is equal to $2p + 1$, where p is the number of the filter poles. This number of the nonzero diagonals of the matrix B for the VK-filter of the first generation can also be designated by $2p + 1$, where $p = 2$.

The formula (6.56) can be calculated in several ways. Inversion and matrix calculations with sparse matrices are available in MATLAB. The second option is to program the calculation in the current programming language. For this option the algorithms will be described. Employing the Cholesky factorisation of the matrix $\mathbf{B}$ into the matrix product $\mathbf{B} = \mathbf{LU}$, where $\mathbf{L}$ is a lower-triangular matrix and $\mathbf{U} = \mathbf{L}^T$ is an upper-triangular matrix, is the easiest way to solve the equation system (6.55). The only condition for the Cholesky factorisation is that all the main minor determinants are equal to a positive value that can be easily proved.

The algorithm for calculation of the upper triangular matrix is shown in Table 6.4. This algorithm starts from the upper left corner of the matrix U and proceeds to calculate the matrix

**Table 6.4** Cholesky–Banachiewicz algorithm.

| Non-zero entries of the matrix $\mathbf{U}$ | Input matrix |
|---|---|

$u_{1,1} = \sqrt{b_{1,1}}$

$j = 2,\dots,p$

$\quad i = 1,\dots,j-1$

$$u_{i,j} = \frac{1}{u_{i,i}}\left(b_{i,j} - \sum_{k=1}^{i-1} u_{k,j} u_{k,i}\right)$$

$$u_{j,j} = \sqrt{b_{j,j} - \sum_{k=1}^{j-1} u_{k,j}^2}$$

$j = p+1,\dots,N$

$\quad i = j-p-1,\dots,j-1$

$$u_{i,j} = \frac{1}{u_{i,i}}\left(b_{i,j} - \sum_{k=j-1-p}^{i-1} u_{k,j} u_{k,i}\right)$$

$$u_{j,j} = \sqrt{b_{j,j} - \sum_{k=j-1-p}^{j-1} u_{k,j}^2}$$

*EndDo*

Input matrix

$$\mathbf{B} = \begin{pmatrix} b_{1,1} & b_{1,2} & \cdots & b_{1,p+1} & & \\ b_{1,2} & b_{2,2} & \cdots & b_{2,p+1} & b_{2,p+2} & \\ \cdots & \cdots & \cdots & \cdots & \cdots & \ddots \\ b_{1,p+1} & b_{2,p+1} & \cdots & b_{p+1,p+1} & b_{p+1,p+2} & \ddots \\ & b_{2,p+2} & \cdots & b_{p+1,p+2} & b_{p+2,p+2} & \ddots \\ & & \ddots & \ddots & \ddots & \ddots \end{pmatrix}$$

Output matrix (order of calculations)

$$\mathbf{U} = \begin{pmatrix} u_{1,1} \rightarrow u_{1,2} & \cdots & u_{1,p+1} & & \\ u_{2,2} & \cdots & u_{2,p+1} & u_{2,p+2} & \\ & \cdots & \cdots & \cdots & \ddots \\ & & u_{p+1,p+1} & u_{p+1,p+2} & \ddots \\ & & & u_{p+2,p+2} & \ddots \\ & & & & \ddots & \ddots \end{pmatrix}$$

column by column. As this upper matrix $\mathbf{U}$ is a transpose of the matrix $\mathbf{L}$, the version of the algorithm comes from Cholesky–Banachiewicz.

The main advantage of the Cholesky factorisation algorithm is that it saves the number of the nonzero diagonals in the triangular matrices at the value $p+1$. The solution of the system (6.55) is broken down into two linear equation systems, the forward reduction and backward substitution.

| Forward reduction (first system) | Backward substitution (second system) | |
|---|---|---|
| $z_1 = y_1/u_{1,1}$ | $x_N = z_1/u_{N,N}$ | (6.57) |
| $z_2 = \left(y_2 - u_{1,2}z_1\right)/u_{2,2}$ | $x_{N-2} = \left(z_{N-2} - u_{N-1,N}x_N\right)/u_{N-1,N-1}$ | (6.58) |
| .... | .... | |
| $j = p+1,\dots,N$ | $j = N-(p+1),\dots,1$ | (6.59) |
| $z_j = \left(y_j - u_{j-1,j}z_{j-1} \dots - u_{j-p,j}z_{j-p}\right)/u_{j,j}$ | $x_j = \left(z_j - u_{j,j+1}x_{j+1} \dots - u_{j,j+p}x_{j+p}\right)/u_{j,j}$ | (6.60) |

In the forward reduction, the linear equation system $\mathbf{L}\mathbf{z} = \mathbf{y}$ ($\mathbf{U}^T\mathbf{z} = \mathbf{y}$) for an unknown vector $\mathbf{z}$ is solved while in the backward substitution the unknown vector $\mathbf{x}$ of the equation system $\mathbf{U}\mathbf{x} = \mathbf{z}$ is evaluated.

The theory of symmetric matrices of the $\mathbf{A}^T\mathbf{A}$ type shows that the positive definiteness ensures that a small positive constant is added to the elements on the main diagonal. The unit may seem a relatively small constant in comparison with the value of $r^2\mathbf{A}^T\mathbf{A}$. The value of weighting factor $r$ has to be limited not to lose the effect of adding unity to the main matrix diagonal on the positive definiteness by rounding the diagonal elements due to the limited bit number for saving numbers in a computer memory. The symmetric matrix of the $\mathbf{A}^T\mathbf{A}$ type for the VK filter of the first generation is as follows

$$\mathbf{A}^T\mathbf{A} = \begin{bmatrix} 1 & -c_1 & 1 & & & \\ -c_1 & (c_1^2+1) & -(c_1+c_2) & 1 & & \\ 1 & -(c_1+c_2) & (c_2^2+2) & -(c_2+c_3) & 1 & \\ & 1 & -(c_2+c_3) & (c_3^2+2) & -(c_3+c_4) & \cdots \\ & & 1 & -(c_3+c_4) & (c_4^2+2) & \cdots \\ & & & \cdots & \cdots & \cdots \end{bmatrix} \qquad (6.61)$$

Because the absolute value of the variable $c_n$ is less than or equal to 2, the maximum value of the elements of the main diagonal of the matrix $\mathbf{A}^T\mathbf{A}$ is equal to 6.

The symmetric matrices of the $\mathbf{A}^T\mathbf{A}$ type for the VK filter of the second generation are shown in Table 6.5. The elements on the main diagonal stabilise at constant values except the first few and last few columns and rows. The steady-state value of the element on the main

**Table 6.5** Matrices of the $\mathbf{A}^T\mathbf{A}$ type.

1-pole filter:

$$\begin{pmatrix} 1 & -1 & & \\ -1 & 2 & -1 & \\ & -1 & 2 & \ddots \\ & & \ddots & \ddots \end{pmatrix}$$

2-pole filter:

$$\begin{pmatrix} 1 & -2 & 1 & & \\ -2 & 5 & -4 & 1 & \\ 1 & -4 & 6 & -4 & \ddots \\ & 1 & -4 & 6 & \ddots \\ & & \ddots & \ddots & \ddots \end{pmatrix}$$

3-pole filter:

$$\begin{pmatrix} 1 & -3 & 3 & -1 & & \\ -3 & 10 & -12 & 6 & -1 & \\ 3 & -12 & 19 & -15 & 6 & \ddots \\ -1 & 6 & -15 & 20 & -15 & \ddots \\ & -1 & 6 & -15 & 20 & \ddots \\ & & \ddots & \ddots & \ddots & \ddots \end{pmatrix}$$

4-pole filter:

$$\begin{pmatrix} 1 & -4 & 6 & -4 & 1 & & \\ -4 & 12 & -28 & 22 & -8 & 1 & \\ 6 & -28 & 53 & -52 & 28 & -8 & \ddots \\ -4 & 22 & -52 & 69 & -56 & 28 & \ddots \\ 1 & -8 & 28 & -56 & 70 & -56 & \ddots \\ & 1 & -8 & 28 & -56 & 70 & \ddots \\ & & \ddots & \ddots & \ddots & \ddots & \ddots \end{pmatrix}$$

**Table 6.6**  The maximal value of the weighting factor.

| Pole number | $p = 1$ | $p = 2$ | $p = 3$ | $p = 4$ |
|---|---|---|---|---|
| $\left(r^2 \mathbf{A}^T \mathbf{A}\right)_{i,i}$ | $2r^2$ | $6r^2$ | $20r^2$ | $70r^2$ |
| $r_{MAX} \approx$ | $7 \times 10^6$ | $4 \times 10^6$ | $2 \times 10^6$ | $1.1 \times 10^6$ |

diagonal for the first-generation of the VK filter and the second generation of the VK 2-pole filter is the same.

Taking into consideration the maximal value of the diagonal components of the matrix $r^2 \mathbf{A}^T \mathbf{A}$ and 14 decimal places for double-precision computer-arithmetic, the limit value $r_{MAX}$ for the weighting factor is shown in Table 6.6.

## 6.4.5   The Transfer Function of the VK Filter

As shown in Table 6.6 diagonal elements of the matrix $\mathbf{A}^T \mathbf{A}$ are stabilised at a constant value with the exception of a short beginning and end of a record, which is irrelevant in the case where the record contains thousands of samples. The number of the first and last elements of the main diagonal whose values differ from the middle section is not greater than the number of the filter poles $p$. Other diagonals have a shorter transition section than the transient part of the main diagonal. The same applies to the diagonals of the matrix (6.61) if the frequency to be tracked is a constant, $c_n = c$. The same can be said for nonzero diagonals of the upper triangular matrix $\mathbf{U}$.

The VK filters of the first and second generation have some common characteristics. Based on the properties of the triangular matrix $\mathbf{U}$ with exception of the aforementioned short beginning and end section it is evident that the coefficients $u_{j-i,j}$ and $u_{j,j+i}$ for $i = 0, \ldots, p$ in the formulas (6.60) for forward reduction and backward substitution are identical. The steady-state value of the coefficients depends on the difference between the index of the row and column. Therefore, it is possible to simplify the index of these matrix elements

$$u_{j,j} = u_0, \quad u_{j-1,j} = u_{j,j+1} = u_1, \ldots, u_{j-p,j} = u_{j,j+p} = u_p \qquad (6.62)$$

The Eq. (6.60) in the forward reduction is equivalent to the linear digital filter of the order $p$ of the infinite impulse response (IIR) type. The input signal is $y_n, n = 1, \ldots, N$ while the output signal is $z_n, n = 1, \ldots, N$. The transfer function in Z-transform is as follows

$$H_F(z) = \frac{Z(z)}{Y(z)} = \frac{1}{u_0 + u_1 z^{-1} + u_2 z^{-2} + \cdots + u_p z^{-p}} \qquad (6.63)$$

Substituting the term $\exp(j\Omega)$ for the complex quantity $z$ in the transfer function (6.63) where $\Omega = \omega T_S$ for $\Omega \in (-\pi, +\pi)$, we obtain the frequency response of the forward reduction operation

$$H_F\left(e^{j\Omega}\right) = \frac{1}{u_0 + u_1 e^{-j\Omega} + u_2 e^{-j2\Omega} + \cdots + u_p e^{-jp\Omega}} \qquad (6.64)$$

Taking into account the reverse order of the samples $z_N, \ldots, z_1$ in the backward substitution which corresponds to the reverse direction of passage of time and the steady-state values of coefficients (6.62), except the positive exponents of powers of the complex variable $z$ the filtration process is based on the same transfer function as for the forward reduction. The input signal is $z_n, n = 1, \ldots, N$ while the output signal is $x_n, n = 1, \ldots, N$. The transfer function and frequency response are as follows

$$H_B(z) = \frac{Z(z)}{Y(z)} = \frac{1}{u_0 + u_1 z^{+1} + u_2 z^{+2} + \cdots + u_p z^{+p}}$$

$$H_B\left(e^{-j\Omega}\right) = \frac{1}{u_0 + u_1 e^{+j\Omega} + u_2 e^{+j2\Omega} + \cdots + u_p e^{+jp\Omega}}.$$

(6.65)

The frequency responses $H_F\left(e^{+j\Omega}\right)$ and $H_B\left(e^{-j\Omega}\right)$ differ in the sign of the frequencies which are opposite and therefore for magnitudes. For these frequency responses the magnitude is identical $\left|H_F\left(e^{+j\Omega}\right)\right| = \left|H_B\left(e^{-j\Omega}\right)\right|$ and the phase is opposite $\arg\left(H_F\left(e^{+j\Omega}\right)\right) = \arg\left(H_B\left(e^{-j\Omega}\right)\right)$. Altogether the forward reduction and backward substitution results in zero-phase digital filtering analogous to the *filtfilt* function in Matlab. Both filters are connected in series as shown in Figure 6.9. The transfer function of the serial connection of the filters is as follows $H_\Sigma(z) = H_F(z) H_B(z)$.

Due to the opposite phase the frequency response of the VK filter is a real positive function as follows

$$H_\Sigma\left(e^{j\Omega}\right) = \left|H_F\left(e^{+j\Omega}\right)\right| \times \left|H_B\left(e^{-j\Omega}\right)\right| = \left|\frac{1}{u_0 + u_1 e^{-j\Omega} + u_2 e^{-j2\Omega} + \cdots + u_p e^{-jp\Omega}}\right|^2$$

(6.66)

### 6.4.6  Bandwidth of the VK Filter

The properties of the VK-filter as a frequency filter are different depending on the origin of this filter. The filter of the first generation is the type of the bandpass filter while the filter of the second generation is a low-pass filter. The relative bandwidth $\delta f$ of the VK filter will be related to the Nyquist frequency like the normalised frequency ranging from 0 to 1. This type

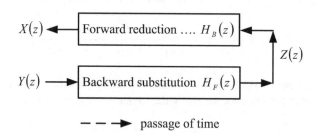

**Figure 6.9**  Forward reduction and backward substitution filters.

of the bandwidth is recommended instead of the absolute bandwidth $\Delta f$ because of the need to emphasise the cut-off frequency of the signal, which is the Nyquist frequency.

The cut-off frequency of a filter is that frequency at which the output power of a signal has dropped to half of its mid-band value. This reduction in a signal level by 6 dB corresponds to the attenuation of the signal amplitude $\sqrt{2}$ times. The 3 dB frequency bandwidth of the digital filter with the transfer function (6.66) may be calculated from the condition.

$$\left| H_\Sigma \left( e^{+j\Omega} \right) \right| = 1 \Big/ \sqrt{2} \tag{6.67}$$

### 6.4.6.1   VK Filter of the First Generation

The frequency response of this second order filter is given by the formula

$$H_F \left( e^{j\Omega} \right) = \frac{1}{u_0 + u_1 e^{-j\Omega} + u_2 e^{-j2\Omega}}. \tag{6.68}$$

By using Euler's formula we can decompose the complex harmonic into the real and imaginary part

$$\exp \left( -j\Omega \right) = \cos \left( \Omega \right) - j \sin \left( \Omega \right). \tag{6.69}$$

Using Euler's formula we get the squared magnitude of the frequency response in the form

$$H_\Sigma \left( e^{j\Omega} \right) = \left| H_F \left( e^{j\Omega} \right) \right|^2 = \frac{1}{u_0^2 + u_1^2 + u_2^2 + 2 \left( u_0 u_1 + u_1 u_2 \right) \cos \left( \Omega \right) + 2 u_0 u_2 \cos \left( 2\Omega \right)}. \tag{6.70}$$

Assuming a constant tracking frequency for which is $c_n = c$ the matrix $\mathbf{B}$ is simplified in the following form

$$\mathbf{B} = r^2 \mathbf{A}^T \mathbf{A} + \mathbf{E} = \begin{bmatrix} r^2 & -r^2 c & r^2 & & & \\ -r^2 c & r^2 \left( c^2 + 1 \right) & -2r^2 c & r^2 & & \\ r^2 & -2r^2 c & r^2 \left( c^2 + 2 \right) & -2r^2 c & r^2 & \\ & r^2 & -2r^2 c & r^2 \left( c^2 + 2 \right) & -2r^2 c & \cdots \\ & & r^2 & -2r^2 c & r^2 \left( c^2 + 2 \right) & \cdots \\ & & & \cdots & & \cdots & \cdots \end{bmatrix} \tag{6.71}$$

The steady-state values of the coefficients on the previous matrix diagonal are as follows

$$b_0 = b_{j,j} = r^2 \left( 2 + c^2 \right) + 1, b_1 = b_{j,j+1} = b_{j,j-1} = -2cr^2, b_2 = b_{j,j+2} = b_{j,j-2} = r^2. \tag{6.72}$$

The Cholesky decomposition gives the following formulas for the coefficients of the denominator of the transfer function

$$
\begin{aligned}
u_2 &= b_2/u_0 & \Rightarrow & \quad b_2 = u_0 u_2 \\
u_1 &= (b_1 - u_1 u_2)/u_0 & \Rightarrow & \quad b_1 = u_0 u_1 + u_1 u_2 \\
u_0 &= \sqrt{b_0 - u_1^2 - u_2^2} & \Rightarrow & \quad b_0 = u_0^2 + u_1^2 + u_2^2.
\end{aligned}
\tag{6.73}
$$

The formula (6.70) is simplified in the following way

$$
H_\Sigma\left(e^{j\Omega}\right) = \left|H_F\left(e^{j\Omega}\right)\right|^2 = \frac{1}{b_0 + 2b_1 \cos(\Omega) + 2b_2 \cos(2\Omega)}
\tag{6.74}
$$

Using the formula (6.72) we get the squared magnitude of the frequency response (6.74) in the form

$$
\frac{1}{b_0 + 2b_1 \cos(\Omega) + 2b_2 \cos(2\Omega)} = \frac{1}{1 + c^2 r^2 - 4cr^2 \cos(\Omega) + 4r^2 \cos^2(\Omega)}.
\tag{6.75}
$$

The transfer function of the VK filter is the unit value for the frequency $\Omega = \omega_C T_S$ at which the tracking filter is tuned. It is valid that $\cos\left(\omega_C T_S\right) = c/2$ and after substituting $\omega_C T_S$ for $\Omega$ in Eq. (6.75) the next formula can be proved

$$
\left|H_F\left(e^{j\omega_C T_S}\right)\right| = \left|\frac{1}{u_0 + u_1 e^{-j\omega_C T_S} + u_2 e^{-2j\omega_C T_S}}\right| = 1
\tag{6.76}
$$

The frequency belongs to the pass band if a harmonic signal of this frequency is attenuated by less than 3 dB $\approx 20\log\left(\sqrt{2}\right)$. The VK filter bandwidth can be determined from the following equation

$$
\frac{1}{1 + c^2 r^2 - 4cr^2 \cos(\Omega) + 4r^2 \cos^2(\Omega)} = \frac{1}{\sqrt{2}}.
\tag{6.77}
$$

The last formula is converted to a quadratic equation in the unknown quantity $\cos(\Omega)$

$$
1 - \sqrt{2} + c^2 r^2 - 4cr^2 \cos(\Omega) + 4r^2 \cos^2(\Omega) = 0.
\tag{6.78}
$$

The quadratic equation has two roots

$$
(\cos(\Omega))_{L,H} = \frac{c}{2} \pm \frac{\sqrt{\sqrt{2}-1}}{2r} = \cos\left(\omega_C T_S\right) \pm \frac{\sqrt{\sqrt{2}-1}}{2r}.
\tag{6.79}
$$

Subscripts $L$ and $H$ denote low-pass (L) and high-pass (H) cut-off frequency. The explicit formula for the angle $\Omega_{L,H} = \omega T_S = 2\pi f T_S$ in radians and or frequency $f_{L,H}$. in hertz is as follows

$$\Omega_{L,H} = \arccos\left(\cos\left(\omega_C T_S\right) \pm \frac{\sqrt{\sqrt{2}-1}}{2r}\right),$$

$$f_{L,H} = \frac{1}{2\pi T_S}\arccos\left(\cos\left(\omega_C T_S\right) \pm \frac{\sqrt{\sqrt{2}-1}}{2r}\right). \tag{6.80}$$

It is appropriate to define the relative bandwidth frequency in the dimensionless normalised frequency that is related to the Nyquist frequency. The relative bandwidth $\delta f$ is as follows

$$\delta f = \frac{\Delta f}{f_S/2} = \frac{f_H - f_L}{f_S/2} \tag{6.81}$$

The relative bandwidth of the first generation VK filter is given by a formula

$$\delta f = \frac{1}{\pi}\left(\arccos\left(\cos\left(\omega_C T_S\right) - \frac{\sqrt{\sqrt{2}-1}}{2r}\right) - \arccos\left(\cos\left(\omega_C T_S\right) + \frac{\sqrt{\sqrt{2}-1}}{2r}\right)\right) \tag{6.82}$$

The simple approximation of the previous formula (6.82) can be obtained by substituting $\omega_C T_S + \Delta\Omega$ for $\Omega$ and simplification of the formula by using an approximation formula for trigonometrics such as $\cos\left(\Delta\Omega\right) \approx 1$ and $\sin\left(\Delta\Omega\right) \approx \Delta\Omega$ for a small value of the angle difference $\Delta\Omega = \Delta\omega T_S$. The result of this simplification is as follows

$$\cos\left(\omega_C T_S + \Delta\Omega\right) = \cos\left(\omega_C T_S\right)\cos\left(\Delta\Omega\right) - \sin\left(\omega_C T_S\right)\sin\left(\Delta\Omega\right)$$

$$\approx \cos\left(\omega_C T_S\right) - \sqrt{1 - \left(\cos\left(\omega_C T_S\right)\right)^2}\,\Delta\Omega. \tag{6.83}$$

The substitution of the approximation (6.83) for the term $\cos\left(\omega_C T_S + \Delta\Omega\right)$ by the quadratic equation (6.78) leads to the formulas

$$\cos\left(\omega_C T_S\right) - \sqrt{1 - \left(\cos\left(\omega_C T_S\right)\right)^2}\,\Delta\Omega_{L,H} \approx \cos\left(\omega_C T_S\right) \pm \frac{\sqrt{\sqrt{2}-1}}{2r}$$

$$\Delta\Omega_{L,H} \approx \mp\frac{1}{\sqrt{1 - \left(\cos\left(\omega_C T_S\right)\right)^2}}\frac{\sqrt{\sqrt{2}-1}}{2r}. \tag{6.84}$$

The absolute bandwidth of the bandpass filter in radians is given by the difference

$$\Delta\Omega = \Delta\Omega_H - \Delta\Omega_L \approx \frac{1}{\sqrt{1 - \left(\cos\left(\omega_C T_S\right)\right)^2}} \frac{\sqrt{\sqrt{2} - 1}}{r}. \tag{6.85}$$

The relative bandwidth of the bandpass filter in the dimensionless quantity $\delta f$ can be approximated for a small value of $\Delta\Omega = \Delta\omega T_S = 2\pi\Delta f T_S$ with the use of the formula

$$\delta f \approx \frac{1}{\pi} \frac{1}{\sqrt{1 - \left(\cos\left(\omega_C T_S\right)\right)^2}} \frac{\sqrt{\sqrt{2} - 1}}{r}. \tag{6.86}$$

The relative bandwidth of the bandpass filter is inversely proportional to the weighting factor $r$ and also depends on the value of $\cos\left(\omega_C T_S\right)$. The zero value of the cosine factor is reached at the tracked frequency equal to half the Nyquist frequency, which means that the relative bandwidth is the narrowest.

If a certain value of the relative bandwidth $\delta f$ is prescribed then the weighting factor $r$ has to be calculated according to the following formula

$$r \approx \frac{1}{\pi \delta f} \frac{\sqrt{\sqrt{2} - 1}}{\sqrt{1 - \left(\cos\left(\omega_C \Delta t\right)\right)^2}}. \tag{6.87}$$

### 6.4.6.2   VK Filter of the Second Generation

The frequency response of this first order filter is given by the formula

$$H_F\left(e^{j\Omega}\right) = \frac{Z\left(e^{j\Omega}\right)}{Y\left(e^{j\Omega}\right)} = \frac{1}{u_0 + u_1 e^{-j\Omega}} \tag{6.88}$$

The transfer function is characteristic for a low-pass filter.

With the use of Euler's formula we get the squared magnitude of the frequency response in the form

$$\frac{1}{\left|u_0 + u_1 e^{-j\Omega}\right|^2} = \frac{1}{\left|u_0 + u_1\left(\cos\left(\Omega\right) - j\sin\left(\Omega\right)\right)\right|^2} = \frac{1}{u_0^2 + u_1^2 + 2u_0 u_1 \cos\left(\Omega\right)}. \tag{6.89}$$

The matrix **B** is of the following shape

$$\mathbf{B} = r^2 \mathbf{A}^T \mathbf{A} + \mathbf{E} = \begin{bmatrix} r^2 + 1 & -r^2 & & \\ -r^2 & 2r^2 + 1 & -r^2 & \\ & -r^2 & 2r^2, 1 & -r^2 \\ & & \cdots & \cdots & \cdots \end{bmatrix} \tag{6.90}$$

The steady-state values of the coefficients on the previous matrix diagonal are as follows

$$b_0 = b_{j,j} = 2r^2 + 1, \quad b_1 = b_{j,j+1} = b_{j,j-1} = -r^2. \tag{6.91}$$

The Cholesky decomposition results in the following formulas for the coefficients of the denominator of the transfer function

$$u_1 = b_1/u_0 \quad \Rightarrow \quad b_1 = u_0 u_1$$
$$u_0 = \sqrt{b_0 - u_1^2} \quad \Rightarrow \quad b_0 = u_0^2 + u_1^2 \tag{6.92}$$

The formula (6.70) is simplified in the following way

$$H_{\Sigma}\left(e^{j\Omega}\right) = \left|H_F\left(e^{j\Omega}\right)\right|^2 = \frac{1}{b_0 + 2b_1 \cos\left(\Omega\right)}. \tag{6.93}$$

Using the formula (6.92) we get the squared magnitude of the frequency response (6.93) in the form

$$\frac{1}{b_0 + 2b_1 \cos\left(\Omega\right)} = \frac{1}{1 + 2r^2 - 2r^2 \cos\left(\Omega\right)}. \tag{6.94}$$

The attenuation within the filter passband which is less than 3 dB can be determined from the following equation

$$\frac{1}{1 + 2r^2 - 2r^2 \cos\left(\Omega\right)} = \frac{1}{\sqrt{2}}. \tag{6.95}$$

This formula is converted to a linear equation for calculation of the unknown quantity $\cos\left(\Omega\right)$

$$1 - \sqrt{2} + 2r^2 \left(1 - \cos\left(\Omega\right)\right) = 0. \tag{6.96}$$

The relative bandwidth frequency in the dimensionless normalised frequency $\delta f$ is as follows

$$\delta f = \frac{\Delta f}{f_S/2} = \frac{f_H}{f_S/2} \tag{6.97}$$

**Table 6.7**   Weighting factor.

| Number of poles | Calculation of the weighting factor for a relative bandwidth | Approximation |
|---|---|---|
| 1 | $r = \sqrt{\dfrac{\sqrt{2}-1}{2\left(1-\cos\left(\pi\delta f\right)\right)}}$ | $r \approx \dfrac{0.2048624}{\delta f}$ |
| 2 | $r = \sqrt{\dfrac{\sqrt{2}-1}{6-8\cos\left(\pi\delta f\right)+2\cos\left(2\pi\delta f\right)}}$ | $r \approx \dfrac{0.0652097315}{\delta f^2}$ |
| 3 | $r = \sqrt{\dfrac{\sqrt{2}-1}{20-30\cos\left(\pi\delta f\right)+12\cos\left(2\pi\delta f\right)-2\cos\left(3\pi\delta f\right)}}$ | $r \approx \dfrac{0.020756902}{\delta f^3}$ |

The cut-off frequency of this filter in radians is as follows $\Omega = \pi\delta f$. For the desired relative bandwidth $\delta f$ the weighting factor $r$ can be calculated by the following formula

$$r = \sqrt{\frac{\sqrt{2}-1}{2\left(1-\cos\left(\pi\delta f\right)\right)}}. \tag{6.98}$$

Using the approximating formula for the cosine function of a small angle

$$\cos\left(\pi\delta f\right) \approx \left(1-\left(\pi\delta f\right)^2/2\right), \tag{6.99}$$

the formula (6.98) may be simplified to the form

$$r \approx \frac{0.2048624}{\Delta f}. \tag{6.100}$$

By using the same procedure formulas for filters with a higher number of poles can be derived. The calculation results are summarised in Table 6.7.

Due to rounding errors the size of the weighting factor is limited. For the same reasons, the minimal value of the bandwidth is limited as well. For double-precision calculations, the minimum percentage of the bandwidth is given in Table 6.8.

In cases where the frequency of the tracked order is not equal to a constant value and it is not possible to evaluate the appropriate weighting factor, the loss function (6.54) is transferred to the form $J = \varepsilon^T \mathbf{R}^T \mathbf{R} \varepsilon + \eta^T \eta$, where $\mathbf{R}$ is a square diagonal matrix

$$\mathbf{R} = \text{diag}\left(r_{1,1}, r_{2,2}, \ldots, r_{N-p,N-p}\right) \tag{6.101}$$

**Table 6.8**   The relative percentage bandwidth of the second generation filter.

| Pole number | $p = 1$ | $p = 2$ | $p = 3$ | $p = 4$ |
|---|---|---|---|---|
| $100\delta f >$ | $5 \times 10^{-6}\%$ | $0.025\%$ | $0.5\%$ | $2\%$ |

with nonzero diagonal elements, which are equal to the weighting factors determined for the instantaneous order frequency and the filter bandwidth, for example, $r_{i,i} = r\left(f_C, \delta f\right)$. The VK-filtration can work with both the absolute bandwidth $\Delta f$ in Hz and the relative bandwidth $\Delta f / f_C$ in percentage to remain a constant value.

### 6.4.7 The Frequency Response of the VK Filter

The transfer function of the filters of both the filter generations have been described in the previous subsections. The practical examples of the frequency responses for each filter generation will be listed separately.

#### 6.4.7.1   VK Filter of the First Generation

The magnitude of the frequency response of the VK filter first generation is shown in Figure 6.10. The phase delay of this filter is equal to zero. The following formula was used for calculation

$$H_\Sigma\left(e^{j2\pi f T_S}\right) = \frac{1}{1 + c^2 r^2 - 4cr^2 \cos\left(2\pi f T_S\right) + 4r^2 \cos^2\left(2\pi f T_S\right)}. \tag{6.102}$$

For a large value of the weighting factor the filter bandwidth is very narrow. The filter selectivity is the best one for the frequency equal to half the Nyquist frequency. The low selectivity of the filter is for frequencies below 0.1 and above 0.9 multiples of the Nyquist frequency.

Selecting a large weighting factor slows the rising and falling edge of the output signal as shown in Figure 6.11. For a large weighting factor the changes in amplitude of the tracked

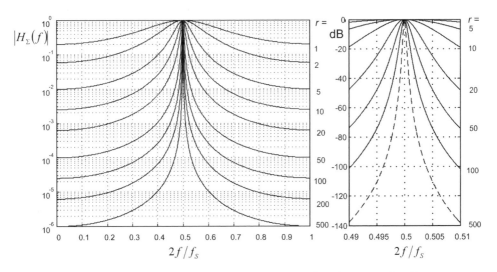

**Figure 6.10**   Magnitude of the frequency response of the first generation filter.

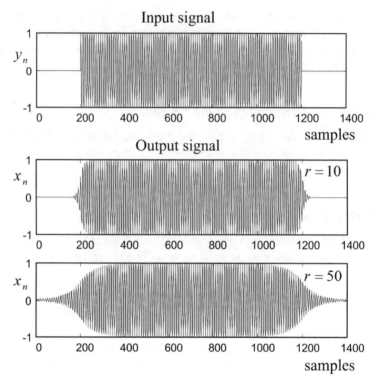

**Figure 6.11**   Rising and falling edge of the output signal VK filter of the second generation.

order are slow. The filter response time $T_R$ is inversely proportional to the absolute bandwidth $\Delta f$ of the bandpass filter. The product of the bandwidth in hertz and the filter response time in seconds is approximately equal to the unit. We can modified this rule in such a way

$$1 = T_R \Delta f = (T_R f_S/2)\left(\frac{\Delta f}{f_S/2}\right) = \frac{T_R}{2T_S}\delta f = \frac{n_R}{2}\delta f$$

$$2 = n_R \delta f$$

(6.103)

where $n_R$ is the number of samples in time $T_R$, for example, for $\delta f = 0.01$ we need 200 samples to settle the filter output.

In this section only the second generation filter with two poles will be described as an example. The magnitude of the frequency response of this filter is shown in the panel on the left side of Figure 6.12. The phase delay of this filter is also equal to zero. The following formula was used for calculation

$$H_\Sigma\left(e^{j2\pi fT_S}\right) = \frac{1}{1 + 6r^2 - 8r^2\cos\left(2\pi fT_S\right) + 2r^2\cos\left(2\pi fT_S\right)}.$$

(6.104)

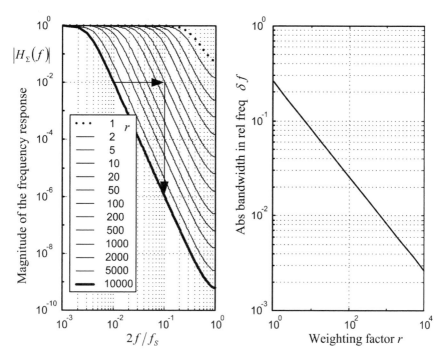

**Figure 6.12**   Magnitude of the frequency response of the second generation filter with two poles.

The dependence of the value of the relative bandwidth on the value of the weighting factor is shown in the panel on the right side of Figure 6.12. The relative bandwidth of the filter is 0.03% of the Nyquist frequency. The graph of the frequency response has a typical shape containing a cut-off point and a slope section with the roll-off of −40 dB per decade. The roll-off of the second generation filter is equal to −20$p$ dB per decade, where $p$ is the number of the filter poles.

To demonstrate features of the VK filter of the second generation the bandpass filter of the CPB (Constant Percentage Band) signal analyser LabShop PULSE is chosen as a reference. According to the technical documentation the 1/3-octave band filter uses the 6-pole filter and complies with the IEC 225-1966, DIN 45651 and ANSI S1.11-1986, Order 3 Type 1-D. The tolerance band is limiting frequency response of the filter at the top (H-Limit) and bottom (L-Limit) and determines the features of filters. The VK filters of the second generation which are referred to in this book use from one to four poles. Due to the double filtration the VK filter has twice the number of poles, that is, from two to eight poles.

An example in Figure 6.13 shows the effect of the VK filter which is tuned to the centre frequency of 100 Hz on the swept sinusoidal signal of the frequency ranging from 0.5 to 1.5 multiple of this centre frequency. The filter bandwidth was set to 23%, which corresponds to 1/3-octave filter. The result of comparison of frequency responses of the VK filter to the tolerance limits of 1/3-octave filter is shown in Figure 6.13. The tolerance limits for the VK filters in the passband are unnecessarily broad. Only the 1-pole filter falls outside the tolerance zone. The 2-pole filter may not meet tolerances for frequencies outside the shown

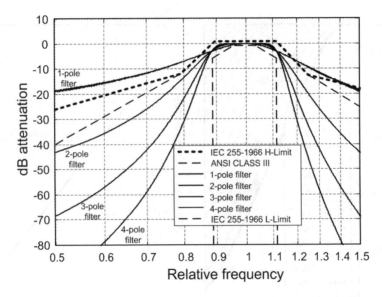

**Figure 6.13**  Comparison of the VK filters with the 1/3-octave 6-pole filter.

interval. The 3-pole and 4-pole filters are fully compliant with the tolerance zone of the
1/3-octave filter.

### 6.4.8  Global Solution for the Multiorder Tracking Filtration

The chapter deals with an algorithm of multiorder tracking of more than one harmonic component. We will focus only on the second generation filters. There are two methods for solution
of the multiorder tracking filtration without beating interference. The first method is based on
the analytical approach and the second one uses the PCG algorithm that combines the direct
and iterative solution [8, 12]. The analytical approach allows only decoupling of two crossing
orders, that is, two harmonic signals which are differing in frequency except a certain time
moment when both components can have the same frequency. In contrast to the analytical
approach the PCG method is a tool for more than the two orders.

In this chapter, we assume that the output of the VK filter is composed of $P$ signals that
represent the indexed envelopes $\mathbf{x}_k, k = 1, \ldots, P$. These unknown are minimising the loss
function

$$J = \sum_{k=1}^{P} r_k^2 \boldsymbol{\varepsilon}_k^T \boldsymbol{\varepsilon}_k + \boldsymbol{\eta}^T \boldsymbol{\eta} = \sum_{k=1}^{P} r_k^2 \mathbf{x}_k^H \mathbf{A}^T \mathbf{A} \mathbf{x}_k + \left( \mathbf{y}^T - \sum_{k=1}^{P} \mathbf{x}_k^H \mathbf{C}_k^H \right) \left( \mathbf{y} - \sum_{k=1}^{P} \mathbf{C}_k \mathbf{x}_k \right), \quad (6.105)$$

where $r_k$ is a parameter determining the bandwidth of the VK filter and $\mathbf{C}_k$ is a diagonal matrix.

$$\mathbf{C}_k = \operatorname{diag} \left\{ \exp \left( j\Theta_k (1) \right), \ldots, \exp \left( j\Theta_k (N) \right) \right\}. \quad (6.106)$$

To determine the loss function's minimum the first derivative with respect to the unknown vectors has to be evaluated. After resetting the derivative to zero the unknown envelopes are obtained as a solution of the following system of equations

$$\frac{\partial J}{\partial \mathbf{x}_k^H} = \mathbf{B}_k \mathbf{x}_k + \mathbf{C}_k^H \sum_{\substack{i=1 \\ i \neq k}}^{P} \mathbf{C}_i \mathbf{x}_i - \mathbf{C}_k^H \mathbf{y} = \mathbf{0}, \quad k = 1, \dots, P, \qquad (6.107)$$

where the following substitutions $\mathbf{B}_k = r_k^2 \mathbf{A}^T \mathbf{A} + \mathbf{E}$ are used and $\mathbf{C}_i^H \mathbf{C}_i = \mathbf{E}$ is a identity matrix. The properties of the symmetric positive definite matrices then $\mathbf{B}_k$ were described above. We only notice that the matrices $\mathbf{B}_k$ may not be identical due to the value of $r_k$ which depends on the order frequency and the bandwidth of the bandpass filter. Unlike the derivation of the filtration algorithm which results in only one signal, apart from indexing the vector $\mathbf{x}_k$ we also need index the matrices $\mathbf{B}_k$ and $\mathbf{C}_k$. The index which equals to 1 was omitted above in matrices $\mathbf{B}_1$ and $\mathbf{C}_1$.

The direct solution of the system of (6.107) is based on calculation of the matrix inverse but an iterative method is more suitable due to the large size of matrices and vectors. For these purposes the definition of matrix $\mathbf{B}$ will be extended in comparison to the chapter dealing with the single order tracking. The matrix $\mathbf{B}$, the vector $\mathbf{x}$ of the unknown variables and the vector $\mathbf{b}$ have the following structure

$$\mathbf{B} = \begin{pmatrix} \mathbf{B}_1 & \mathbf{C}_1^H \mathbf{C}_2 & \cdots & \mathbf{C}_1^H \mathbf{C}_P \\ \mathbf{C}_2^H \mathbf{C}_1 & \mathbf{B}_2 & \cdots & \mathbf{C}_2^H \mathbf{C}_P \\ \cdots & \cdots & \cdots & \cdots \\ \mathbf{C}_P^H \mathbf{C}_1 & \mathbf{C}_P^H \mathbf{C}_2 & \cdots & \mathbf{B}_P \end{pmatrix}, \quad \mathbf{x} = \begin{pmatrix} \mathbf{x}_1 \\ \mathbf{x}_2 \\ \cdots \\ \mathbf{x}_P \end{pmatrix}, \quad \mathbf{b} = \begin{pmatrix} \mathbf{C}_1^H \mathbf{y} \\ \mathbf{C}_2^H \mathbf{y} \\ \cdots \\ \mathbf{C}_P^H \mathbf{y} \end{pmatrix} \qquad (6.108)$$

The size of the matrix $\mathbf{B}$ is $NP \times NP$ and the size of the vectors $\mathbf{x}$ and $\mathbf{b}$ is $NP \times 1$.

The banded matrices $\mathbf{B}_i$ with the real entries are symmetric and positive definite. For the complex diagonal matrix $\mathbf{C}_i^H \mathbf{C}_j$ it is valid that $\mathbf{C}_i^H \mathbf{C}_j = \left( \mathbf{C}_j^H \mathbf{C}_i \right)^H$, therefore $\mathbf{B}$ is a Hermitian matrix system of the equations (6.107). The structure of the nonzero entries of the matrix with the four blocks $P = 4$ is shown in Figure 6.14. Individual blocks are square matrices with the number of lines reaching tens or hundreds of thousands. The block matrices on the main diagonal are the banded matrices of the $(2m + 1)$ bandwidth which are also called sparse matrices. The block matrices outside of the main diagonal are the diagonal matrices.

For the matrix which is composed from more than two blocks it is almost impossible to find an explicit formula for solving the system of equations. It is much better to solve this system of equations using an iterative algorithm. A monograph describing the iterative solution of the system of linear equations [13] recommend the conjugate gradient method with a preconditioner matrix which is called the Preconditioned Conjugate Gradient (PCG) algorithm which is particularly suitable for solving the linear equations with the sparse symmetric and positively definite (SPD) matrices.

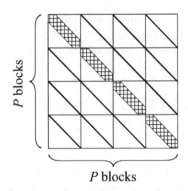

Figure 6.14   A block matrix composed of sparse matrices.

### 6.4.9   Analytical Solution for Decoupling of Two Components

The analytical solution is possible only for decoupling two components (orders). Of course it makes sense to solve this problem when the frequency components are the same at only one time. If $P = 2$ then the equations (6.108) are simplified to

$$
\mathbf{B} = \begin{pmatrix} \mathbf{B}_1 & \mathbf{C}_1^H \mathbf{C}_2 \\ \mathbf{C}_2^H \mathbf{C}_1 & \mathbf{B}_2 \end{pmatrix}, \quad \mathbf{x} = \begin{pmatrix} \mathbf{x}_1 \\ \mathbf{x}_2 \end{pmatrix}, \quad \mathbf{b} = \begin{pmatrix} \mathbf{C}_1^H \mathbf{y} \\ \mathbf{C}_2^H \mathbf{y} \end{pmatrix} \tag{6.109}
$$

where $\mathbf{C}_1$ and $\mathbf{C}_2$ are diagonal matrices which are defined by (6.106).
   The direct solution of the blockwise system (6.109) in the form of the sub vectors $\mathbf{x} = \mathbf{B}^{-1}\mathbf{b}$ is as follows [9]

$$
\begin{aligned}
\mathbf{x}_1 &= \left( \mathbf{B}_2 \mathbf{C}_2^H \mathbf{C}_1 \mathbf{B}_1 - \mathbf{C}_2^H \mathbf{C}_1 \right)^{-1} \left( \mathbf{B}_2 - \mathbf{E} \right) \mathbf{C}_2^H \mathbf{y} \\
\mathbf{x}_2 &= \left( \mathbf{B}_1 \mathbf{C}_1^H \mathbf{C}_2 \mathbf{B}_2 - \mathbf{C}_1^H \mathbf{C}_2 \right)^{-1} \left( \mathbf{B}_1 - \mathbf{E} \right) \mathbf{C}_1^H \mathbf{y}.
\end{aligned} \tag{6.110}
$$

### 6.4.10   Iterative Methods of Solution

The PCG method is a combination of the conjugate gradient (CG) method and the technique based on using the preconditioner matrix. The CG-method was first described by authors Hestenes, M.R. & E. Stiefel in 1952 [14]. The outline of the principle of the CG-method is described in the monograph which was written by Saad [6]. First, however, it is defined by the inner product of two complex vectors $\mathbf{u}$ and $\mathbf{v}$ in which the bar over a variable represents the complex conjugate

$$
(\mathbf{u}, \mathbf{v}) = \sum_{i=1}^{N} u_i \bar{v}_i. \tag{6.111}
$$

If $(\mathbf{u}, \mathbf{v}) = 0$ then the vectors $\mathbf{u}$ and $\mathbf{v}$ are called orthogonal. Furthermore, the vectors $\mathbf{u}$ and $\mathbf{v}$ are called as $\mathbf{B}$-conjugate if it is valid $(\mathbf{Bu}, \mathbf{v}) = 0$. The last expression means orthogonality of vectors $\mathbf{Bu}$ and $\mathbf{v}$.

For a description of the calculation procedure we assume that the problem is to solve the system of equations $\mathbf{Bx} = \mathbf{b}$ in which the sparse matrix $\mathbf{B}$ of the $NP \times NP$ size is Hermitian ($\mathbf{B} = \mathbf{B}^H$) and positively definite $\mathbf{x}^T\mathbf{Bx} > 0$ for all nonzero vectors $\mathbf{x}$ of the $NP \times 1$ size. As was mentioned before, due to the large size of the matrix $\mathbf{B}$, the calculation, which is based on calculation of the inverse of this matrix, is not technically possible. It is therefore necessary to propose a solution that uses only successive approximations of the solution, which contain only a vector-matrix multiplication without the matrix inverse. The iterative search for the solution in the $NP$-dimensional space starts from an initial guess $\mathbf{x}_0$. For the CG-method the next direction of the search along a line in the $NP$-dimensional space and the previous one are $\mathbf{B}$-conjugated.

Solving the mentioned linear system of equations using the CG-method is equivalent to searching for a minimum of the real loss function

$$f(\mathbf{x}) = \frac{1}{2}\mathbf{x}^T\mathbf{Bx} - \mathbf{x}^T\mathbf{b}. \tag{6.112}$$

The gradient of this loss function with respect to the vector of unknown $\mathbf{x}$ is as follows

$$\nabla_{\mathbf{x}}f(\mathbf{x}) = \frac{\partial f(\mathbf{x})}{\partial \mathbf{x}} = \mathbf{Bx} - \mathbf{b}. \tag{6.113}$$

The zero gradient (6.113) corresponds to the minimum of the loss function (6.112). The negative gradient is a residual which is defined by $\mathbf{r} = \mathbf{b} - \mathbf{Bx}$.

According to the CG-method a new estimate of the vector $\mathbf{x}$ using an iterative process in the $m$-th step can be written as

$$\mathbf{x}_{m+1} = \mathbf{x}_m + \alpha_m\mathbf{p}_m, \tag{6.114}$$

where $\alpha_m$ is a scalar parameter which modifies the length of the vector $\mathbf{p}_m$ which determines the direction in which the CG-method tries to find a new approximation of the solution.

If the residual vector in the $m$-th iteration step of searching for solution of $\mathbf{Bx} = \mathbf{b}$ is marked by

$$\mathbf{r}_m = \mathbf{b} - \mathbf{Bx}_m \tag{6.115}$$

then the new residual vector in the next iteration step with respect to (6.115) is given by the formula

$$\mathbf{r}_{m+1} = \mathbf{b} - \mathbf{Bx}_{m+1} = \mathbf{b} - \mathbf{B}\left(\mathbf{x}_m + \alpha_m\mathbf{p}_m\right) = \mathbf{r}_m - \alpha_m\mathbf{Bp}_m. \tag{6.116}$$

The CG-method requires that the residual vectors $\mathbf{r}_m$ and the previous one $\mathbf{r}_m - \alpha_m\mathbf{Bp}_m$ are orthogonal, that is, $\left(\mathbf{r}_m - \alpha_m\mathbf{Bp}_m, \mathbf{r}_m\right) = 0$, then it can be derived that

$$\alpha_m = \frac{\left(\mathbf{r}_m, \mathbf{r}_m\right)}{\left(\mathbf{Bp}_m, \mathbf{r}_m\right)}. \tag{6.117}$$

The next search direction of the refined approximation can be expressed as a linear combination of the previous direction of finding solutions and the new value of the residual vector

$$\mathbf{p}_{m+1} = \mathbf{r}_{m+1} + \beta_m\mathbf{p}_m, \tag{6.118}$$

where $\beta_m$ is a scalar parameter. As was stated before the consecutive directions of the search is to be **B**-conjugated, therefore, vector $\mathbf{Bp}_m$ is orthogonal to vector $\mathbf{p}_{m-1}$. The consequence of the above equation is as follows

$$\begin{aligned}
\left(\mathbf{Bp}_m, \mathbf{r}_m\right) &= \left(\mathbf{Bp}_m, \mathbf{p}_m - \beta_{m-1}\mathbf{p}_{m-1}\right) \\
&= \left(\mathbf{Bp}_m, \mathbf{p}_m\right) - \beta_{m-1}\left(\mathbf{Bp}_m, \mathbf{p}_{m-1}\right) = \left(\mathbf{Bp}_m, \mathbf{p}_m\right).
\end{aligned} \tag{6.119}$$

The orthogonality condition $\left(\mathbf{p}_{m+1}, \mathbf{Bp}_m\right) = 0$ for the successive directions of the search result in the following equation

$$\left(\mathbf{p}_{m+1}, \mathbf{Bp}_m\right) = \left(\mathbf{r}_{m+1} + \beta_m\mathbf{p}_m, \mathbf{Bp}_m\right) = \left(\mathbf{r}_{m+1}, \mathbf{Bp}_m\right) + \beta_m\left(\mathbf{p}_m, \mathbf{Bp}_m\right) = 0. \tag{6.120}$$

Equation (6.120) implies that the scalar parameter $\beta_m$ is given by the formula

$$\beta_m = \frac{\left(\mathbf{r}_{m+1}, \mathbf{Bp}_m\right)}{\left(\mathbf{p}_m, \mathbf{Bp}_m\right)}. \tag{6.121}$$

Since we can express the expression $\mathbf{Bp}_m$ according to (6.116)

$$\mathbf{Bp}_m = \frac{1}{\alpha_m}\left(\mathbf{r}_{m+1} - \mathbf{r}_m\right) \tag{6.122}$$

then the parameter $\beta_m$ can be calculated using the formula

$$\beta_m = \frac{1}{\alpha_m}\frac{\left(\mathbf{r}_{m+1}, \left(\mathbf{r}_{m+1} - \mathbf{r}_m\right)\right)}{\left(\mathbf{Bp}_m, \mathbf{r}_m\right)} = \frac{\left(\mathbf{r}_{m+1}, \mathbf{r}_{m+1}\right)}{\left(\mathbf{r}_m, \mathbf{r}_m\right)}. \tag{6.123}$$

The final version of the CG algorithm according to Saad's monograph [6] is shown in Table 6.9 as a CG algorithm. The first direction of the search is given by the negative gradient $\mathbf{p}_0 = \mathbf{r}_0$ as is shown in Figure 6.15.

**Table 6.9** Iterative algorithm for solution a large system of linear equations.

| CG algorithm | PCG algorithm, version 1 | PCG algorithm, version 2 |
|---|---|---|
| $\mathbf{r}_0 = \mathbf{b} - \mathbf{B}\mathbf{x}_0$ | $\mathbf{r}_0 = \mathbf{b} - \mathbf{B}\mathbf{x}_0$ | $\mathbf{r}_0 = \mathbf{b} - \mathbf{B}\mathbf{x}_0$ |
| $\mathbf{p}_0 = \mathbf{r}_0$ | $\mathbf{z}_0 = \mathbf{M}^{-1}\mathbf{r}_0$ | $\tilde{\mathbf{r}}_0 = \mathbf{L}^{-1}\mathbf{r}_0$ |
| | $\mathbf{p}_0 = \mathbf{z}_0$ | $\mathbf{p}_0 = \mathbf{L}^{-T}\tilde{\mathbf{r}}_0$ |
| $m = 0,1,\ldots,$ until convergence | $m = 0,1,\ldots,$ until convergence | $m = 0,1,\ldots,$ until convergence |
| $\alpha_m = (\mathbf{r}_m, \mathbf{r}_m)/(\mathbf{B}\mathbf{p}_m, \mathbf{p}_m)$ | $\alpha_m = (\mathbf{r}_m, \mathbf{z}_m)/(\mathbf{B}\mathbf{p}_m, \mathbf{p}_m)$ | $\alpha_m = (\tilde{\mathbf{r}}_m, \tilde{\mathbf{r}}_m)/(\mathbf{B}\mathbf{p}_m, \mathbf{p}_m)$ |
| $\mathbf{x}_{m+1} = \mathbf{x}_m + \alpha_m \mathbf{p}_m$ | $\mathbf{x}_{m+1} = \mathbf{x}_m + \alpha_m \mathbf{p}_m$ | $\mathbf{x}_{m+1} = \mathbf{x}_m + \alpha_m \mathbf{p}_m$ |
| $\mathbf{r}_{m+1} = \mathbf{r}_m - \alpha_m \mathbf{B}\mathbf{p}_m$ | $\mathbf{r}_{m+1} = \mathbf{r}_m - \alpha_m \mathbf{B}\mathbf{p}_m$ | $\tilde{\mathbf{r}}_{m+1} = \tilde{\mathbf{r}}_m - \alpha_m \mathbf{L}^{-1}\mathbf{B}\mathbf{p}_m$ |
| $\beta_m = (\mathbf{r}_{m+1}, \mathbf{r}_{m+1})/(\mathbf{r}_m, \mathbf{r}_m)$ | $\mathbf{z}_{m+1} = \mathbf{M}^{-1}\mathbf{r}_{m+1}$ | $\beta_m = (\tilde{\mathbf{r}}_{m+1}, \tilde{\mathbf{r}}_{m+1})/(\tilde{\mathbf{r}}_m, \tilde{\mathbf{r}}_m)$ |
| $\mathbf{p}_{m+1} = \mathbf{r}_{m+1} + \beta_m \mathbf{p}_m$ | $\beta_m = (\mathbf{r}_{m+1}, \mathbf{z}_{m+1})/(\mathbf{r}_m, \mathbf{z}_m)$ | $\mathbf{p}_{m+1} = \mathbf{r}_{m+1} + \beta_m \mathbf{p}_m$ |
| $EndDo$ | $\mathbf{p}_{m+1} = \mathbf{z}_{m+1} + \beta_m \mathbf{p}_m$ | $EndDo$ |
| | $EndDo$ | |

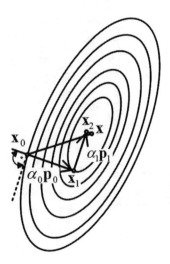

**Figure 6.15**  Iteration procedure for solving the system of equations.

### 6.4.11  PCG Iterative Method

Introduction of the preconditioner matrix $\mathbf{M}$ which suitably approximates matrix $\mathbf{A}$ and the general system of equations $\mathbf{Bx} = \mathbf{b}$ can be used as an easy direct solution of the alternative system $\mathbf{Mx} = \mathbf{u}$ assuming easy inversion of the preconditioner matrix $\mathbf{M}$. As already mentioned, the CG method solves a system of equations with a matrix system, which is symmetric and positive definite. This property is also expected for matrix $\mathbf{M}$. The iterative method is suitable to solve a system of equations

$$\mathbf{BM}^{-1}\mathbf{u} = \mathbf{b}. \tag{6.124}$$

The preconditioner system $\mathbf{Mx} = \mathbf{u}$ enables a better starting guess to be found which is based on the replacement of the matrix $\mathbf{B}$ by the matrix $\mathbf{M}$.

The first variant of improvement of the CG method to the PCG method results in speeding up the calculation that is based on the possibility of easy inversion of the matrix $\mathbf{M}$. A new direction of finding better approximation of solutions using the formula (6.118) is a linear combination of the residual vector $\mathbf{z}_m$, which is the solution preconditioner system, that is, $\mathbf{z}_m = \mathbf{M}^{-1}\mathbf{r}_m$ and an original direction of searching for the previous approximation of the solution. The algorithm of calculation is shown in Table 6.9 as a PCG algorithm, version 1.

The second variant of the PCG method is based on the Cholesky decomposition of the preconditioner matrix into a product of two triangular matrices

$$\mathbf{M} = \mathbf{LL}^T \tag{6.125}$$

For this variant of choosing the preconditioner matrix it is necessary to define the following auxiliary vectors

$$\tilde{\mathbf{p}}_m = \mathbf{L}^T\mathbf{p}_m, \quad \mathbf{u}_m = \mathbf{L}^T\mathbf{x}_m, \quad \tilde{\mathbf{r}}_m = \mathbf{L}^T\mathbf{z}_m = \mathbf{L}^{-1}\mathbf{r}_m, \quad \tilde{\mathbf{A}} = \mathbf{L}^{-1}\mathbf{BL}^{-T}. \tag{6.126}$$

It is possible to deduce that

$$
\begin{aligned}
\left(\mathbf{r}_m, \mathbf{z}_m\right) &= \left(\mathbf{r}_m, \mathbf{L}^{-T}\mathbf{L}^{-1}\mathbf{r}_m\right) = \left(\mathbf{L}^{-1}\mathbf{r}_m, \mathbf{L}^{-1}\mathbf{r}_m\right) = \left(\tilde{\mathbf{r}}_m, \tilde{\mathbf{r}}_m\right) \\
\left(\mathbf{B}\mathbf{p}_m, \mathbf{p}_m\right) &= \left(\mathbf{B}\mathbf{L}^{-T}\tilde{\mathbf{p}}_m, \mathbf{L}^{-T}\tilde{\mathbf{p}}_m\right) = \left(\mathbf{L}^{-1}\mathbf{B}\mathbf{L}^{-T}\tilde{\mathbf{p}}_m, \tilde{\mathbf{p}}_m\right) = \left(\tilde{\mathbf{B}}\tilde{\mathbf{p}}_m, \tilde{\mathbf{p}}_m\right).
\end{aligned}
\tag{6.127}
$$

The previous algorithm can be rewritten in the following sequence

$$
\begin{aligned}
\alpha_m &= \left(\tilde{\mathbf{r}}_m, \tilde{\mathbf{r}}_m\right)/\left(\tilde{\mathbf{B}}\tilde{\mathbf{p}}_m, \tilde{\mathbf{p}}_m\right), \mathbf{u}_{m+1} = \mathbf{u}_m + \alpha_m\tilde{\mathbf{p}}_m \\
\tilde{\mathbf{r}}_{m+1} &= \tilde{\mathbf{r}}_m - \alpha_m\tilde{\mathbf{B}}\tilde{\mathbf{p}}_m, \beta_m = \left(\tilde{\mathbf{r}}_{m+1}, \tilde{\mathbf{r}}_{m+1}\right)/\left(\tilde{\mathbf{r}}_m, \tilde{\mathbf{r}}_m\right) \\
\tilde{\mathbf{p}}_{m+1} &= \tilde{\mathbf{r}}_{m+1} + \beta_m\tilde{\mathbf{p}}_m.
\end{aligned}
\tag{6.128}
$$

The CG algorithm is applied to the preconditioner system of equations

$$
\tilde{\mathbf{B}}\mathbf{u} = \mathbf{L}^{-1}\mathbf{b}
\tag{6.129}
$$

The algorithm of calculation is shown in Table 6.9 as a PCG algorithm, version 2.

### 6.4.12  Initial Guess for the Iterative Solution

The basic trick for the fast convergence of the iterative calculation is that the following preconditioner matrix was used for the solution of the next practical examples

$$
\mathbf{M} = \begin{pmatrix} \mathbf{B}_1 & \mathbf{0} & \cdots & \mathbf{0} \\ \mathbf{0} & \mathbf{B}_2 & \cdots & \mathbf{0} \\ \cdots & \cdots & \cdots & \cdots \\ \mathbf{0} & \mathbf{0} & \cdots & \mathbf{B}_P \end{pmatrix}, \quad
\mathbf{M}^{-1} = \begin{pmatrix} \mathbf{B}_1^{-1} & \mathbf{0} & \cdots & \mathbf{0} \\ \mathbf{0} & \mathbf{B}_2^{-1} & \cdots & \mathbf{0} \\ \cdots & \cdots & \cdots & \cdots \\ \mathbf{0} & \mathbf{0} & \cdots & \mathbf{B}_P^{-1} \end{pmatrix}.
\tag{6.130}
$$

The above blockwise preconditioner matrix can be easily inverted. The advantage of this choice of the preconditioner matrix $\mathbf{M}$ is that the result of calculation is the first guess $\mathbf{x}_0$ of the solution in the form of the separate envelopes for each component (orders) assuming that there is no problem with decoupling of the crossing orders. Therefore, we get

$$
\mathbf{x}_0 = \mathbf{M}^{-1}\mathbf{b}.
\tag{6.131}
$$

The individual envelopes are given by a set of solutions

$$
\mathbf{x}_{0,i} = \mathbf{B}_i^{-1}\mathbf{C}_i^H\mathbf{y}, \quad i = 1, \ldots, P.
\tag{6.132}
$$

We assume that the frequencies of the sinusoidal components differ from each other, but there is a time instant in which the frequencies as a function of time can coincide with each other. The envelopes in these time instants are calculated by using the iterative process.

Decoupling of crossing orders will be demonstrated using an example of superposing sinusoids. It is assumed that the signal which is sampled at 1 kHz is composed of two

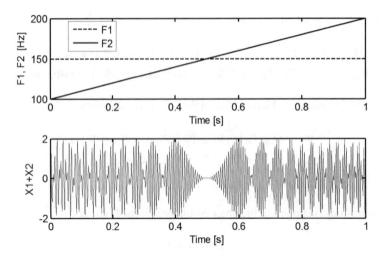

**Figure 6.16**   Superposing two sinusoidal signals.

harmonic components, so called orders, of the same amplitude which is equal to unity. One of these two components has a constant frequency of 150 Hz and the second one has a frequency linearly increasing from 100 to 200 Hz. The resulting signal and instantaneous frequencies as a function of time are shown in Figure 6.16. The frequency of the tracked components is the same in the middle of the time interval. We say that orders are crossing.

In the first step the VK filter tracks the component with increasing frequency (F2). The result of filtration with the use of the first generation filter and the second generation with one to four poles is shown in Figure 6.17. In both cases, the bandpass filters were set to the bandwidth of 10 Hz. Filtering is ineffective at the time of crossing orders.

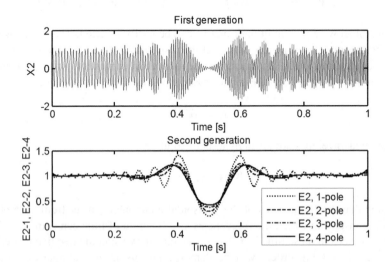

**Figure 6.17**   Result of filtration without decoupling orders.

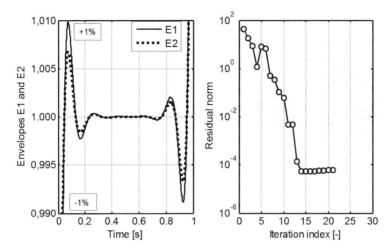

**Figure 6.18**  Envelopes and iterative process as a function of index.

The result of decoupling the crossing orders is shown in Figure 6.18. The amplitude of the two components as a result of the calculation differs from the true amplitude by 1 percent. The variable RELRES is the relative residue NORM (b – B∗x) / NORM (b). To calculate the envelopes the PCG method which is implemented in Matlab was used. The left diagram shows the results of calculation with the use of the PCG algorithm originating from MATLAB.

Theoretically, the solution can be obtained after reaching the number of iteration steps which is equal to the dimension of the unknown vector. The result of decoupling two crossing orders in Figure 6.16 is shown in Figure 6.18.Quite an accurate solution is obtained after 15 iteration steps.

## 6.4.13   Comparison of the VK Filter of the First and Second Generation

The use of tracking filters in analysis or diagnostics of rotary or reciprocating machines requires knowing the rotational speed as a function of time and the frequencies of orders are needed for tuning the filter when tracking them. The rotational speed for each sample of the measured signal are usually interpolated from measurements which provide only the average value over a period of time, such as one complete revolution of a shaft. The value of the orders may be found on the basis of either the machine design or by measuring the frequency spectrum, which identifies the frequency of the dominant components. These dominant components may be due to either the mechanical resonance of certain parts of the machine structure or an excitation whose frequency depends on the rotational speed.

As an example of the use of the VK filter of the first and second generation the analysis of the sound which is measured at a tailpipe of a passenger car will be used. The microphone placement was according to the standard, that is, at a distance of 0.5 m in the direction of 45 degrees from the axis of the exhaust pipe. The rotational speed of the car engine crankshaft was determined from the pulse signal connected to the engine control unit. The sound recorder

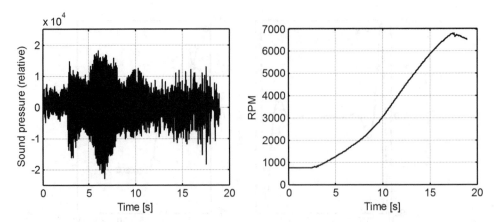

**Figure 6.19**   Sound recorded at the exhaust tailpipe and the engine RPM.

is not in units of sound pressure, because in this demo example, it does not matter. The time history of sound pressure and RPM is shown in Figure 6.19.

The engine SPL at the exhaust tailpipe as a function of time is shown in Figure 6.20. It concerns the average sound pressure in dB with a filter of the fast type. The frequency filter of the A-type was used for calculating SPL. The maximum magnitude of the average sound pressure was taken as a reference for calculating decibels. This option corresponds to the time record of the sound pressure without specific physical units. The graph can determine how many decibels the noise is less than the maximum of SPL. The sound pressure level is increased approximately in proportion to the motor RPM.

The engine is four-stroke and four-cylinder. Twice per revolution of the engine crankshaft the mixture of fuel and air is fired in a cylinder. It is assumed that the even multiples of the engine rotational frequency dominate in the sound spectrum, especially in measurements of the sound pressure near the exhaust tailpipe. Therefore only the even harmonics will be tracked with the use of the VK filter. The largest values of SPL in the speed range up to 4000 RPM can be found for the even harmonics, that is, 2, 4, 6 and 8 ord.

The 2-pole VK filter of the second generation assesses SPL in the frequency range whose frequencies differ from the centre frequency up to the difference of 0.1 ord in relation to the

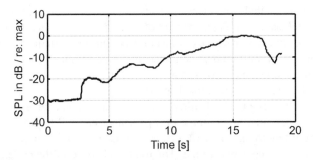

**Figure 6.20**   SPL at the exhaust tailpipe as a function of time.

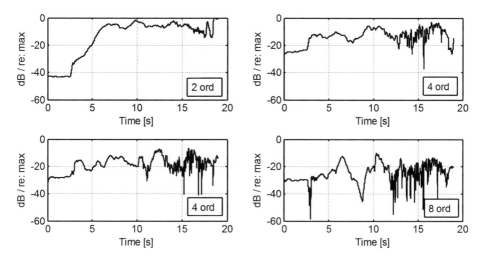

**Figure 6.21**    SPL of the individual orders as a function of time.

frequency to which it is tuned. Therefore the passbands of the VK filter are set from 1.9 to 2.1 ord, then from 3.9 to 4.1 ord, and so on. As a reference for calculating dB the maximum value of the sound pressure at the output of the filter which is tuned to 2 ord is taken. The dependence of SPL of the individual orders on time is shown in Figure 6.21. The diagrams in this figure confirm that the components of the frequency of 2 and 4 ord dominate in the frequency spectrum.

The result of filtration with the VK filter of the first generation is shown in Figure 6.22. The same result can be obtained if we use the envelope as an amplitude modulation signal of the sinusoidal carrier of the same frequency as it is the frequency of the order. The result

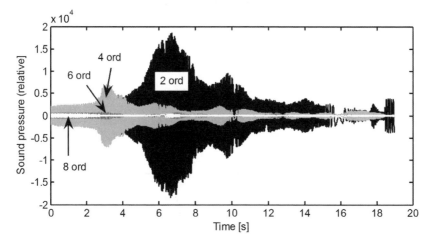

**Figure 6.22**    Output of the first generation of the VK filter.

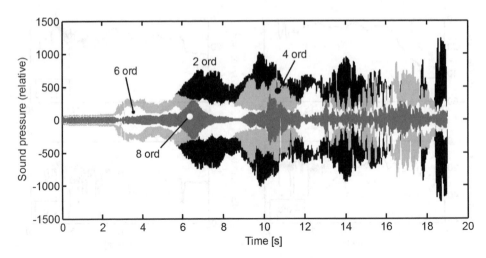

**Figure 6.23** Output of the second generation of the VK filter with orders which are converted into waveforms.

of this calculation in Figure 6.23 is significantly different from the filtration using the first generation filter in Figure 6.22. This difference cen be explained by the frequency responses of both filters. The VK filter of the second generation efficiently attenuates the signal whose frequency is outside the passband range rather than the filter of the first generation of the same bandwidth. The VK filter of the first generation has a very narrow passband which gives good supression of a signal outside the passband frequency range. The passband of the VK filter of the second generation is of well defined shape while the shape of the passband of the first generation filter is similar to the resonance peak and the roll-off (slope) at the sides of the frequency response is not a constant.

The filter of the first generation has a theoretical significance and can be used when the frequency of order is known very accurately. The filter of the second generation allows certain errors in the estimation of the frequency in question.

## References

[1] Gade, S. Herlufsen, H., Konstantin-Hansen, H. and Wismer, N.J. (1995) Order tracking analysis. *Brüel & Kjaer Technical Review*, No. 1.

[2] Gade, S., Herlufsen, H., Konstantin-Hansen, H. and Vold, H. (1999) Characteristics of the Vold-Kalman Order Tracking Filter, *Brüel &Kjaer Technical Review*, 1–1999. Kalman Order Tracking Filter, *B&K Technical Review* No 1 - 1999.

[3] Welch, G. and Bishop, G. An Introduction to the Kalman Filter, Dept of Computer Science University of North Carolina at Chapel Hill Chapel Hill, NC 27599-3175, September 17, 1997. http://www.cs.unc.edu/~welch/kalman/.

[4] Haykin, S. (1996) *Adaptive Filter Theory*, Prentice Hall International.

[5] Dossing, O. (2005) Rotating machinery considerations on tachosignal and shaft speed estimation. Proceedings of the 6th SVIB Vibration Symposium in Rigsgransen, Sweden.

[6] Vold, H. and Leuridan, J. (1993) Order Tracking at Extreme Slow Rates, Using Kalman Tracking Filters. SAE Paper Number 931288.

[7] Gade, S., Herlufsen, H., Konstantin-Hansen, H. and Vold, H. (1999) Characteristics of the Vold/Kalman Order Tracking Filter. *Sound and Vibration*, April 1999. Available from World Wide Web <http://www.ramsete.com/Public/Kalman/S%26Vib_April1999_pp2-8.pdf.

[8] Feldbauer, Ch. and Holdrich, R. (2000) Realisation of a Vold-Kalman Tracking Filter – A Least Square Problem. Proceedings of the COST G-6 Conference on Digital Audio Effects (DAFX-000, Verona Italy, December 7–9.

[9] Tůma, J. (2003) Vold-Kalman filtration in MATLAB (in Czech). Proceedings of Eleventh MATLAB Conference, Humusoft Praha, Praha, 25 November 2003, pp. 575–586.

[10] Tůma, J. (2004) Dedopplerisation in Vehicle External Noise Measurements. Proceedings of Eleventh International Congress on Sound and Vibration, ICSVII, St. Petersburg, 5–8 July 2004, pp. 151–158.

[11] Tůma, J. (2005) Setting the passband width in the Vold-Kalman order tracking filter. Twelfth International Congress on Sound and Vibration (ICSV12), Lisbon, July 11–14, 2005, Paper 719.

[12] Tůma, J. (2013) Algorithms for the Vold-Kalman multiorder tracking filter. 14th International Carpathian Control Conference ICCC 2013, May 26–28, 2013, Rytro, Poland, p. 076.

[13] Saad, Y. (2000) *Iterative Methods for Sparse Linear Systems*, 2nd edn, SIAM.

[14] Hestenes, M.R. and Stiefel, E. (1952) Methods of conjugate gradients for solving linear systems. *Journal of Research of the National Bureau of Standards*, **49**, 409–436.

# 7

# Reducing Noise of Automobile Transmissions

This chapter deals with the practical methods for reducing the noise of vehicle transmissions. This chapter does not give instructions on how to design a quiet gearbox, but describes how to make the right decisions when choosing options. The first topic is focused on the problem of how to predict the effect of changes in the toothing design on the sound pressure level (SPL) which is radiated by the gearbox. The method of measurement of a transmission error, which is closely related to the excitation of parametric vibrations, is described at the beginning of the chapter. For making decisions on noise abatement measures it is important to know how the gearbox contributes to the overall pass-by noise level which is an objective criteria for permission to operate vehicles on public roads. Finally, measures to reduce noise which is radiated by the transmission unit of a truck will be described in the case study. Noise of the truck transmissions is a more serious problem compared to passenger cars, whose gearbox is hidden in the car body.

Besides keeping the objective value of SPL under the required limit the subjective perception of noise still dominates but this is a different subject matter: psychoacoustics.

## 7.1 Normal Probability Plot

New products are rarely tested on a large series of identical prototypes. There is a small amount of data therefore the statistical set is very small and it is impossible to draw a histogram or histograms don't have much visual presence. The appropriate analytical tool is an experimental distribution function (cumulative histogram) in transformed coordinates. This graphical technique enables of quick assessment of whether or not a data set is approximately normally distributed. The data are plotted against a theoretical normal distribution in such a way that the points should form an approximate straight line as can be seen in Figure 7.1. Deviations from this straight line indicate deviations from the normal data set. The normal probability plot is formed by the vertical axis (probability) and horizontal axis with equally

*Vehicle Gearbox Noise and Vibration: Measurement, Signal Analysis, Signal Processing and Noise Reduction Measures*, First Edition. Jiří Tůma.
© 2014 John Wiley & Sons, Ltd. Published 2014 by John Wiley & Sons, Ltd.

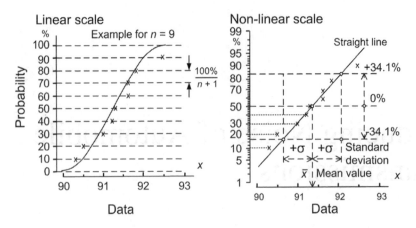

**Figure 7.1**    An empirical cumulative distribution function.

probable values. A mean value which is identical with the median for normally distributed data corresponds to the probability of 50%. The range $50 \pm 34.1\%$ of values around the mean value which is increased and decreased by the standard deviation $\sigma$ is also shown in this diagram. It can be concluded that the greater the slope of the straight line, the smaller the standard deviation. The slope of this line thus at first sight determines whether the statistical data have the same standard deviation.

Steps for constructing a Normal Probability Plot:

- Step 1: Download from Internet (http://www.weibull.com/GPaper/) probability paper for the normal probability plot.
- Step 2: Order the data from smallest to largest.
- Step 3: Number the ordered values $i = 1, 2, \ldots, n$ where the smallest value is numbered 1 and the largest is numbered $n$.
- Step 4: Compute the cumulative probability $p_i = i/(n + 1) \times 100\%$ for $i = 1, 2, \ldots, n$.
- Step 5: Plot the pairs, (ordered value, $p_i$-value), with the measurement scale along the horizontal axis and the $p_i$-scale along the vertical axis.

## 7.2    Transmission Error Measurements

Noise and vibration problems in gearing are mainly concerned with the smoothness of the drive. The parameter that is employed to measure smoothness is the Transmission Error 19 (TE) [1]. This parameter can be expressed as a linear displacement at a base circle radius defined by the difference of the output gear's position from where it would be if the gear teeth were perfect and infinitely stiff. Many references have attested to the fact that a major goal in reducing gear noise is to reduce the transmission error of a gear set. Experiments [2] show that decreasing TE by 10 dB (approximately 3 times less) results in decreasing transmission sound level by 7 dB. The TE of the TATRA gearbox was measured later by using the method developed at the VSB – Technical University of Ostrava [3].

The basic formula for calculating TE of a simple gear set is given as

$$TE = r_2\Theta_2 - r_1\Theta_1 = \left(\Theta_2 - \frac{n_1}{n_2}\Theta_1\right)r_2 \tag{7.1}$$

where $n_1$ and $n_2$ are teeth numbers of pinion and wheel respectively, $\Theta_1$ and $\Theta_2$ are angles of rotation of the mentioned gear pair and $r_1$ and $r_2$ are radiuses of the pinion and wheel, respectively.

TE results not only from manufacturing inaccuracies, such as profile errors, tooth pitch errors and run-out, but from bad design. The pure tooth involute deflects under load due to the finite mesh stiffness caused by tooth deflection. A gear case and shaft system deflects due to load, as well. While running under load, one of the important parameters, tooth contact stiffness, varies, which excites the parametric vibration and, consequently, noise.

TE results from angular vibrations of the meshing gears. There are many possible methods for angular vibration measurement during rotation:

- Tangentially mounted accelerometers,
- Laser torsional vibration metre based on the Doppler effect.
- Incremental rotary encoders (several hundreds of pulses per revolution).

Instantaneous angular velocity is proportional to the reciprocal value of the time interval, which is elapsed between consecutive pulses. The measurement methods for the length of the time interval are as follows:

- Sample number and interpolation.
- High frequency oscillator (100 MHz) and impulse counter.
- Phase demodulation.

The simplest method for evaluation of the instantaneous rotational speed is the reciprocal value of the time interval between two consecutive pulses. If the pulse signal is sampled then the time interval between the adjacent pulses is determined by interpolation some 50 times more accurately than indicated by the actual sampling interval. The accuracy is satisfying for the RPM measurement based on only one pulse per shaft rotation. This method is not suitable if the large number of pulses per revolution is generated, which results in a few samples between pulses and the time interval length is impossible to estimate with any degree of accuracy. If the string of encoder pulses as an analogue signal controls a gate for the high frequency clock signal (100 MHz), which is an input of an pulse counter, then this method works properly. This principle is implemented in the signal analysers produced by Rotec as was mentioned before. The instantaneous angular velocity is primary information for TE evaluation and needs numerical integration with respect to time. Henriksson and Pärssinen present an example of this measurement [4]. The methods based on using Hilbert transform gives as primary information the instantaneous rotation angle. The complexity of measuring TE is also given by the need to synchronise the separate measurements of the individual rotation angle of the two gears in mesh which are the input data for calculation of the time-varying difference between the lengths of two arcs. As is evident many possible approaches exist for measuring

TE, but as Derek Smith states in his book [5], in practice, measurements based on the use of sensors dominates.

## 7.2.1  Averaged Transmission Error for a Circular Pitch Rotation

An example of TE measurement using the encoders, shown in this subchapter, deals with the gear train consisting of the 27- and 44-tooth gears under test. The pinion rotates at 1038 RPM and the moment of the force (torque) about the pinion axis reached 1300 nm. The pinion and wheel are mounted on the countershaft and the output shaft, respectively. The incrementary rotary encoders E1 and E2 are attached to the pinion and wheel shaft as described in the subsection on the measurement of angular vibrations using encoders. Both of the encoders generate a string of pulses. As a consequence of Shannon's sampling theorem, a few pulses have to be recorded during each mesh cycle. This means that the number of pulses produced per encoder revolution must be a multiple of the tooth number. If five harmonics of toothmeshing frequency are required then the number of pulses per gear revolution must be at least ten times greater than the number of teeth. The encoder generating 500 pulses per revolution seems to be an optimum.

The gear speed variation at the toothmeshing frequency results in the phase modulation of the base frequency of the pulse signal which is generated by the encoder. As noted many times before the phase-modulated signal contains sideband components around the carrying component situated in Figure 7.2 at $500 \pm 27k$ orders for the 27-tooth pinion and at $500 \pm 44k$ orders for the 44-tooth wheel, where $k = 1, 2, 3, \ldots$ is an integer. The spectrum frequency axis in Figure 7.2 is in ord. Take note of the fact that the dominating components in both the sideband families exceed the background noise level at least by 20 dB or even more. Both the spectra were evaluated from time signals that are a result of the synchronised averaging of 100 revolutions of gears under testing.

Records corresponding to the complete rotation of the pinion and wheel are input signals for calculation of the Hilbert transform and unwrapped phase. The normalised phases and order spectra of the angular vibration of the countershaft and output shaft including the 27-tooth

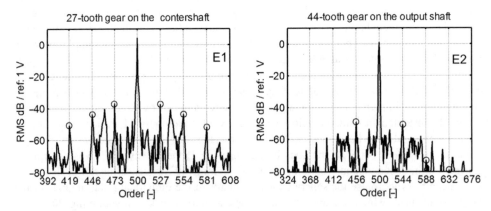

**Figure 7.2**   Order spectrum of phase modulated pulse signal generated by the envoder E1 and E2.

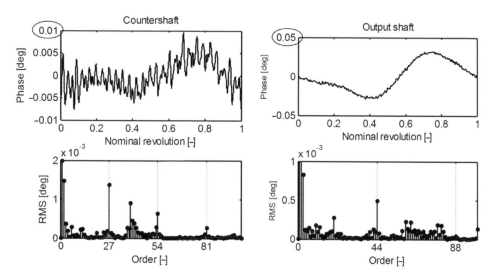

**Figure 7.3**   Angular vibrations of the countershaft and output shaft.

and 44-tooth gears and other gears under load are shown in Figure 7.3. Comparing both the diagrams shows the range of the uniformity rotation.

Angular frequency vibrations at the toothmeshing frequency of the gear train of the 27- and 44-teeth can be obtained with the use of the comb filter through the frequency domain. The centre frequencies of the pass bands of this comb filter are tuned to the harmonics of the toothmeshing frequency. The number of sideband components for the pinion and the wheel are chosen as 3 and 5 respectively. The result of the bandpass filtering is shown in Figure 7.4.

The number of waves in both the diagrams in Figure 7.4 differs and the time delay between these two measurements is unknown. The period corresponding to one rotation by the tooth circular pitch can become a base for the calculation of the transmission error. As in the case of the average toothmesh of the vibration signal, both the phase variations can be averaged again to obtain an averaged representative for angular vibration during the tooth circular pitch

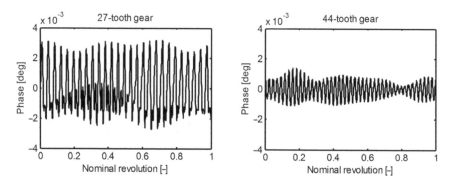

**Figure 7.4**   Angular vibrations of the 27-tooth and 44-tooth gears at their toothmeshing frequencies.

rotation. In this case, the signal is composed only from the harmonics of the toothmeshing frequency. The modulation effect is removed and purely periodic signals are obtained with 27 or 44 identical waves. One of them can be considered as a representative of the time history of angular vibration.

As both the encoder signals are recorded separately the true phase delay between these signals is to be detected. This problem is solved thanks to the fact that the average toothmesh responses, for example in acceleration of some point on the gear case, to dynamic forces acting between meshing teeth are theoretically of the same shape and occur at the same time. The average toothmesh signal which is synthesised in this way can be designated as the second stage of averaging [3, 6]. If both the pulse signals at the output of the encoders are sampled together with the acceleration signal then a way exists of how to align the phase of the pulse signals at least in the toothmeshing period. The lag for the maximum correlation between the originally measured average toothmesh gives the value of the relative delay between these two records. The second stage of averaging can also be used for both angular vibration signals during one complete revolution of the corresponding gears. The second stage of averaging can also be used for both angular vibration signals during one complete revolution of the corresponding gears including aligning their time delay. The result of aligning is shown in Figure 7.5.

The phase of the angular vibration is the same as the angle of rotation which can be transformed into the arc length. The difference between these two lengths of the arc yields TE as a function of circular tooth pitch rotation. The result according to the formula (7.1) for three times repeating circular tooth pitch rotations is shown in Figure 7.6. The three periods of TE were chosen to emphasise periodicity of the result.

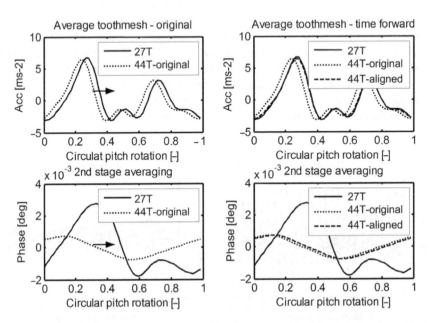

**Figure 7.5**  Compensation of the phase shift between the time course of angular vibrations with the use of eliminating the time delay between the signals at the output of the sensors E1 and E2.

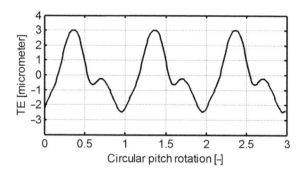

**Figure 7.6**   Three periods of the transmission error as a function of circular tooth pitch rotation.

Another example in Figure 7.7 deals with the measurement of TE to verify modification of the tooth surface. This measurement is related to the analysis of dependance of TE on the moment of force acting at the gear with 27 teeth. Only the higher load shows a small increase of TE and consequently a slightly higher noise level emitted by the gearbox. Note that the transmission error is almost independent of the moment of force, which means that the tooth surface modification is optimal. The design of the TATRA truck gearbox results in TE, which is within the range corresponding to car gearboxes. As Derek Smith claims, the truck gears work with TE in the range of +/− 10 microns [5].

Transmission error is calculated only for an angle of rotation by one circular tooth pitch.and thus characterises the design of teeth of both gears. The continuous measurement of TE during the revolution of the gears is not possible.

## 7.2.2   Transmission Error Measurements during Many Revolutions of Gears

This measurement method is based on calculating the unwrapped phase of the pulse signals and the difference between the phases with respect to the gear ratio [6]. The use of this measurement method of the transmission error is verified by the same gearbox of the same serial number as in previous chapter. The gear train which connects the countershaft and the

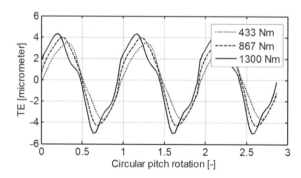

**Figure 7.7**   Transmission error as a function of circular tooth pitch rotation and the moment of force.

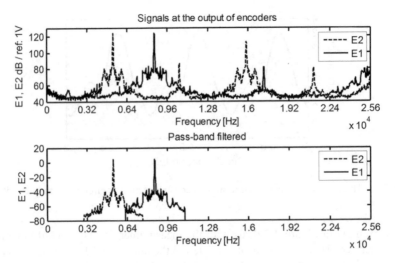

**Figure 7.8**   Frequency spectra of signals at the output of the encoders.

output shaft consists of the 27-tooth pinion and the 44-tooth wheel as in the previous example. The operating conditions related to RPM and the moment of force (torque) were also the same as in the cases above. The record of the acceleration signal is not needed in this case because the signals at the output of the encoders are measured simultaneously. The frequency spectra of the encoder signals are shown in Figure 7.8.

The frequency spectrum of the pulse signal is composed of several harmonics of its basic frequency due to the approximately rectangular shape of the pulses. The basic frequency of pulses at the output of the encoders differs from each other and is given by the product of the number of pulses per revolution and the rotational frequency of the corresponding shaft. To filter out the higher harmonics with the use of a bandpass filter is the first step of the calculations. This filter creates a phase-modulated sinusoidal signal. The result of the bandpass filtration is shown at the bottom part of Figure 7.8. It is possible to employ a digital pass-band filter or the filtration in the frequency domain using the direct and inverse FFT which is preferred. The pass band of the signal generated by the encoder of E1 is centred about the frequency of 8608 Hz and the lower and upper sideband of the bandwidth of 2866 Hz. The properties of the pass band of the second filter which affects the frequency spectrum of the signal generated by the encoder of E2 are as follows: the centre frequency of 5280 Hz and both the sidebands of the width of 1758 Hz. The bandwidth of the sidebands provides a sufficient frequency range for at least four harmonic components of the toothmeshing frequency.

Individual unwrapped phases as a function of time are shown on the left side in Figure 7.9. The phase difference between the two unwrapped phases is on the right side in this figure. The calculation of the phase difference takes into consideration the gearing ratio

$$\Delta\Theta = \left(\Theta_2 - \frac{n_1}{n_2}\Theta_1\right) \tag{7.2}$$

where $n_1 = 27$ and $n_2 = 44$.

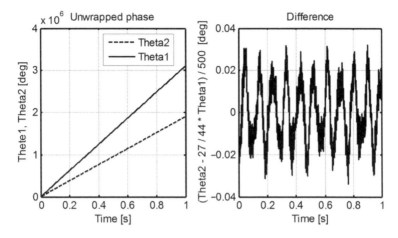

**Figure 7.9**   Unwrapped phase and the difference between them.

The signal of the phase difference contains significant low-frequency components and high-frequency components. As the high-frequency components relate to the noise of the transmission unit the spectrum analysis is focused on them. The average rotational speed in RPM during a complete shaft rotation is not ideal, but fluctuated in the small range as shown in Figure 7.10. The origin of the low frequency component of about 10 Hz could be explained by the low frequency error of the IRC sensor which was analysed at the end of Chapter 5. The rotational frequency of the 44-tooth gear is about 10 Hz and the dominating frequency component of the phase difference is also about 10 Hz. Therefore the frequency of this dominating component is equal to 1 ord with respect to the rotational frequency of the IRC sensor. As is evident from Figure 5.45 the error reaches the magnitude of 0.01 degrees. The topic of the book, however, concerns the audible noise and vibration.

The transmission error is calculated as a product of the phase difference and the radius of the appropriate gear. The frequency spectrum of this difference is shown in Figure 7.11. Only harmonic components of the toothmeshing frequency are related to the tested gears. The

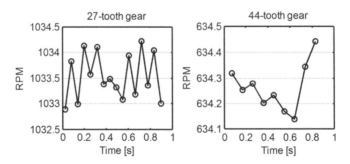

**Figure 7.10**   Variations in the rotational speed of the countershaft and output shaft.

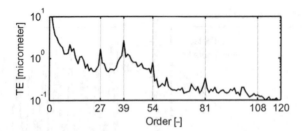

**Figure 7.11**   Frequency spectrum of the weighted phase difference.

spectrum component of the frequency of 39 ord is caused by a loaded driving gear belonging to the other gear train so it will be filtered out.

There are many ways of presenting the transmission error. One way to illustrate this function of time for the entire duration of the measurement is shown in Figure 7.12. The result of the synchronous filtration during a complete revolution of the gear is shown in Figure 7.13 and finally the average toothmesh is shown in Figure 7.14.

Since the plots in Figures 7.5 and 7.13 do not differ much, both measurement methods provide almost the same result. The differences are almost negligible. The common recording of signals at the output of the encoders offers the ability to track the response of changes in torque and determine the average stiffness of the tooth contact.

## 7.3   Case Study

### 7.3.1   Historical Notes

The author of this book was a member of the team for gearbox noise reduction at TATRA, a Czech company that produces off-road and on-road trucks. TATRA trucks are used to carry goods and have a maximum authorised total mass exceeding 12 tonnes. In the past, TATRA truck gearboxes were noted for having a very robust design. Their only disadvantage was that they were 'a little noisy'. The introduction of the 80 dB limit for the peak pass-by noise level

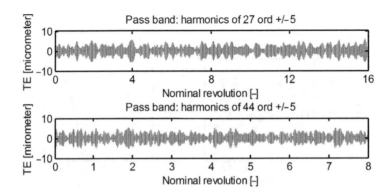

**Figure 7.12**   Transmission error as a function of the gear revolutions.

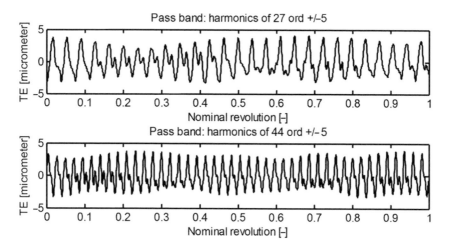

**Figure 7.13**   The averaged transmission error during the gear complete rotation.

of on-road trucks in Europe in 1994 created a challenge to start research works on gearbox noise reduction. It should be noted that the pass-by noise measurements shall be made using the filter of the FAST type and the frequency A-weighting on the dB scale at a distance of 7.5 m from the centre of the test track. In the USA, the SPL is measured at a distance of 50 feet (double that of the European test method), therefore 6 dB has to be added to the USA limit to make it correspond to the European level. The truck's operational condition during pass-by noise measurements is based on the full acceleration test specified in the International Standard, ISO R362-1982 (E) 'Acoustic-Measurement of Noise Emitted by Accelerating Road Vehicles-Engineering Methods' or SAE J366 Surface Vehicle standard – (R) 'Exterior Sound Level for Heavy Trucks and Buses (Issued 1969-07, Reaffirmed 1987-02)'. The test requires the vehicle to be driven through a test track at full acceleration (see Figure 7.15).

The noise level of road vehicles was decreased in intervals as it was possible to fulfil the requirements given by the vehicle noise legislation. The peak SPL of the mentioned category of trucks was set to 84 dB before 1994. The 84 dB trucks were easy to produce but the last

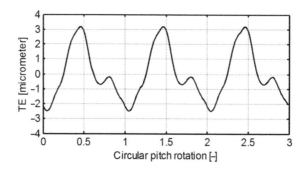

**Figure 7.14**   Three periods of the averaged transmission error as a function of circular pitch rotation.

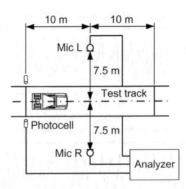

**Figure 7.15**   Pass-by vehicle noise measurements (ISO R362).

decrease of the SPL by 4 dB required fundamental improvements. The truck engine is the noisiest unit. Engine noise reduction for TATRA trucks of the T 815-2 type, introduced into production in 1989, was reached by using an enclosure that fit over the engine compartment and, also, by reducing the rate of the rise in cylinder pressure. The reduction of the SPL to reach the last 4 dB was not easy. To solve the gearbox noise problem it was important to decide if the special enclosure should be extended over the gearbox or not. As mentioned before, the designers decided to solve the gearbox noise problem at the very source. The experiments show that the effect of the enclosure, having an opening for the Cardan drive shafts, on the SPL, is less than the 3 dB reduction at a distance of 1 m. The TATRA truck power train has the gearbox separated from the engine and these units are connected to it by the Cardan shaft, therefore, it is not easy to design the enclosure protecting noise radiation of the gearbox. This chapter describes the research work on silencing the truck gearbox. As previously stated, the truck gearbox is without enclosure and its operating rotational speed and load are not steady but are variable. The experience gained from research work on the truck gearbox noise reduction can be applied, generally, to any other transmission.

### 7.3.2   Vehicle Pass-by Noise Measurements

Transmission noise should be assessed in terms of external vehicle noise. In order to analyse the contributions of each individual simple-gear train to the overall noise level, it is preferable to measure the time history of sound pressure and engine revolutions per minute (RPM) during the pass-by test. The vehicle transmission units contain multiple shafts that may run coherently through fixed transmissions, or partially related through torque converter slippage, or independently, as in, for instance, a cooling fan in an engine compartment. The emitted tonal noise consists of harmonic components whose frequency is a multiple of the corresponding part's rotational speed (RPM). It is supposed that the instantaneous rotational speed is measured in the form of a pulse train and is then transmitted via a radio transmitter to be recorded together with the noise and threshold signals. The overall pass-by vehicle noise level is almost given by the contributions of the mentioned tonal components. Therefore, it is possible to analyse contributions of each individual simple gear train as well as the engine firing noise and the tyre noise to the overall noise level. The noise spectrum analysis requires extending the measurement instrumentation by introducing methods to calculate the instantaneous engine RPM during pass-by tests. In contrast to the measurement with a stationary vehicle, during a pass-by noise test, the effect of the relative speed of the noise source and the microphones is translated into a Doppler phenomenon and causes positive (approaching) or negative (receding) frequency shifts (typically from 2 to 5%) in the signals received by the microphones. The frequency $f_0$ of sound waves that are emitted by a source moving at velocity $v_R$ relative to the stationary microphone receiving it, shifts the sound waves frequency to the value of $f_1$:

$$f_1 = \frac{c}{c + v_r} f_0 \tag{7.3}$$

where $c$ is the sound velocity. The Doppler frequency shift can lead to considerable errors for order tracking using a narrow- band pass filter.

The analysis of the truck pass-by noise sources based on the frequency spectra decomposition was introduced at TATRA in 1994 (first published in 1996) [7]. An example of the results is shown in Figure 7.16 and contains a plot of the relative sound power contributions (root mean square (RMS) squared) versus time elapsed from the moment when the truck crosses the test track entry point. The data corresponds to gearbox improvements discussed later. It is assumed that all the shafts run coherently through fixed transmissions. Each power contribution is the sum of five partial contributions associated with corresponding harmonics of a base frequency. The base frequency and five of its harmonics of the sound waves excited

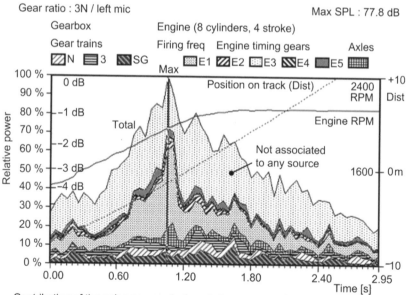

Contribution of the noise sources in the relative power to the maximum of Total:
1961 RPM, 8.6 m, N: 1.9%, 3: 1.8%, SG: 6.7%.     Gearbox: 10%, Axles: 8.6%,
E1: 41%, E2: 4.7%, E3: 2.5%, E4: 0.3%, E5: 4.5%.  Engine: 54%, Background: 27%.

**Figure 7.16**  Truck noise sources contribution to overall pass-by noise level.

by the truck's 8-cylinder, 4-stroke engine divided into the engine firing frequency (E1) and timing gears (E2 through E5), main gearbox gears (gear trains marked by N, 3) and drop gearbox (SG), axle gears (Axles), and eventually tyre profile were evaluated using the engine rotational speed measurement (RPM) after correction according to Eq. (7.3) and prediction of the truck's instantaneous velocity and position (Dist) on the track, taking into account the Doppler shift and the 1% relative passband width around the centre frequency. Note the cursor values at the peak value (Max) of the sound pressure level. The frequency range of sound pressure was 3200 Hz and the line number was 800 therefore the spectrum averages signals over time intervals of 250 ms.

### 7.3.3 Estimation of the Doppler Frequency Shift

The distance $x$ of a vehicle, running at velocity $v_V$, from the beginning of the test track is given by the following formula (see Figure 7.18)

$$x = \int_0^t v_V dt_1 \qquad (7.4)$$

The vehicle velocity depends on the driving wheel circumferential velocity $v_T$ that is proportional to the wheel rotational speed and wheel dynamic radius. Furthermore the wheel rotational speed is proportional to the engine rotational speed that is supposed to be measured during test and transmitted by wireless equipment to the stationary signal analyser.

As the vehicle accelerates, a slip between the wheel circumference velocity $v_T$ and vehicle velocity $v_V$ can arise

$$s_p = \frac{v_T - v_V}{v_T}. \qquad (7.5)$$

This slip is determined the longitudinal force coefficient that is defined as a ratio $F_T/F_Z$, where $F_T$ is the longitudinal force and $F_Z$ is the normal force acting on the tyre by the road surface. Assuming that the force coefficient does not exceed its maximum value, the dependence of the longitudinal force coefficient on the slip can be modelled in a limited range by a linear function of the tyre slip as shown in Figure 7.17.

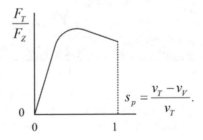

**Figure 7.17** Longitudinal force coefficient vs. slip.

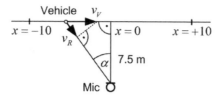

**Figure 7.18**   Velocity and position of a vehicle on the test track.

It follows thence that the vehicle acceleration is proportional to the longitudinal tyre slip and so it is possible to evaluate the vehicle's instantaneous velocity $v_V$ using Euler's method for the numeric solution of the following differential equation

$$\frac{d^2x}{dt^2} = \frac{dv_V}{dt} = K\frac{v_T - v_V}{v_T} \tag{7.6}$$

where $K$ is a coefficient depending on the longitudinal tyre slip and the normal force acting on the tyre.

The relationship between the vehicle position on the test track and the microphone is shown in Figure 7.18. The noise source's relative velocity to the stationary microphone is given by the formula

$$v_R = v_V \sin(\alpha) \tag{7.7}$$

where $\sin(\alpha) = x\big/\sqrt{x^2 + 7{,}5^2}$.

The derived formula enables the vehicle's instantaneous velocity and position on the test track to be calculated only taking into account the vehicle engine rotational speed. This information about the vehicle motion enables the instantaneous value of the Doppler factor (7.3) that enables a dedopplerisation of the tracking orders to be estimated.

### 7.3.4   Pass-by Noise Analysis with the Use of the Vold-Kalman Filter

The example deals with the pass-by-noise measurement of a heavy-duty vehicle according to the aforementioned ISO standard. The vehicle was driven through the test track at full acceleration. The maximum SPL always fits into the prescribed track of the vehicle. The vehicle producer's aim when measuring the pass-by-noise noise is to predict the outcome of the official ISO test and to get an overview of the intensity of the source of noise which is radiated by the vehicle. The measurements were pretriggered by 250 ms the moment before the vehicle passes the photocells at the beginning of the 20-metre test track. The total recording time of the signal was 6 seconds. The wireless connection transmits a tachosignal which consists of a string of pulses. The frequency of the tachopulses is the same as the frequency of the engine crankshaft rotation. The time history of the noise signal from the left microphone is shown in Figure 7.19. The sampling frequency for all the signals is equal to 8192 Hz, which means that the frequency range of measurement is equal to 3200 Hz.

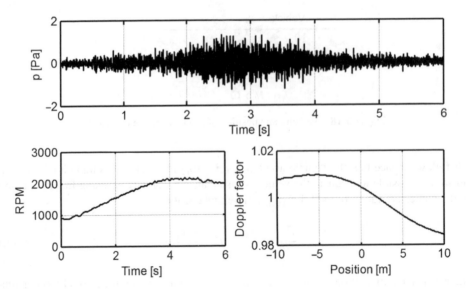

**Figure 7.19** The time course of sound pressure, engine rotational speed and Doppler factor during the test of the truck pass-by noise.

The tachosignal is a source of information about the instantaneous engine rotational speed in RPM. Transmitting the analogue tacho signal wirelessly without delay is highly preferred. The interpolation using the cubic spline curve is shown in Figure 7.19. The vehicle arrives at the beginning of the test track at a steady velocity. The value of the Doppler factor as a function of the vehicle position is shown in Figure 7.19. The reference point of the vehicle where the Doppler ratio is equal to the unit is situated at the front part of its cab. The gearbox as a source of noise is situated at a distance of 2 m from the aforementioned reference point. This fact requires the position of the noise source to be corrected for calculating the Doppler factor. Therefore the Doppler factor is equal to one only when the cab front part moves 2 metres forward from position zero as can be seen in the right panel at the bottom of Figure 7.19.

The Vold-Kalman filter was tuned to the toothmeshing frequency of the 27-tooth gear and its second and third harmonics. The filter bandwidth is required to be equal to 0.1 ord of the engine rotational speed. The bandwidth in percentage for the basic frequency is equal to $0.1/27 \times 100\% = 0.37\%$ while the bandwidth for the second harmonic component is equal to the half of 0.37% that is 0.18%. Both of the bandwidths are very narrow in comparison to the frequency shift due to the Doppler effect. The plot of the noise level corresponding to the first three harmonics of the 27-tooth gear of the vehicle gearbox against the vehicle position on the test track is shown in Figure 7.20.

Regardless of these previous results, software was later developed for pass-by noise analysis based on the Vold-Kalman order-tracking filtration, which included a mathematical model of the tyre slip as a function of the longitudinal and normal force acting on the tyre by the road surface [7]. This is assuming that the tyre slip and the vehicle position at the end of the test track end is corrected by less than 1 m which is negligible compared to the truck dimensions.

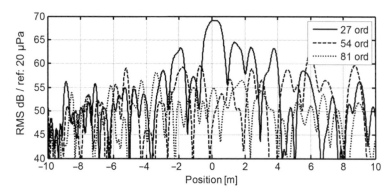

**Figure 7.20**   Sound pressure level of the 27-tooth gear as a fiction of the truck position on the test track.


## 7.3.5   Factory Limit for the Gearbox SPL

To fulfil the pass-by noise level limit for the truck tests according to the aforementioned International Standard the factory internal limit for the gearbox noise level measured on a test stand by microphones located at a distance of 1 m was proposed by technicians having experience with noise problems. This limit level was used to check gearbox quality and as a criterion for improving gearbox design. The gearbox noise was measured on the test stand at the gearbox's operational condition, simulating the vehicle pass-by noise test. The SPL limit was set up according to a comparison of the results of the many pass-by and test stand measurements. The aforementioned SPL limit for gearboxes is a maximum of the overall SPL in the RPM range at the gearbox input shaft corresponding to the engine RPM range during acceleration on the test track (where the pass-by noise tests take place). The reason for this is that the peak vehicle SPL should be less than the required limit for all the tested gears. There are two possible arrangements of the test stand, an open-loop test stand and a back-to-back test rig. In contrast to the open-loop test stand, the back-to-back test rig configuration saves drive energy. The torque to be transmitted by the gearbox is induced by a planetary gearbox. The gearbox under testing is enclosed in a semi-anechoic room. The quality of the semi-anechoic room is of great importance for the accuracy and reliability of the results. According to ISO 3744, the reverberation time should satisfy a condition required for the ratio of the room absorption to the measurement surface area (greater than 6) in the frequency range from at least 200 Hz to 3 kHz. The input shaft speed is slowly increased from minimal to maximal RPM while the gearbox is under a load corresponding to full vehicle 'acceleration' during the pass-by tests. In TATRA, noise is measured by two microphones located by the side of the gearbox under a test at a distance of 1 m. Accelerometers that are attached on the surface of the gearbox housing, near the shaft bearings, can extend information about the noise sources. A tacho probe, that generates a string of pulses, is usually employed to measure the gearbox's primary-shaft (input shaft) rotational speed.

The effect of the gearbox improvement has to be demonstrated statistically. The gearbox design, namely the gear design parameters, and the production quality before the introduction of the noise reduction improvements in 1994 can be described by the empirical distribution function of the tested gearbox noise level. The empirical distribution function was chosen

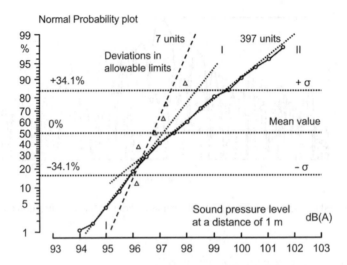

**Figure 7.21**  Empirical cumulative distribution function of sound pressure level of the gearbox on the test rig.

because of the low number of the tested specimens for which the set of the statistical data is also small and it is impossible to construct a histogram.

As was explained before, the gearbox SPL is measured during the run-up tests on the test stand as a maximum of overall SPL in the RPM range corresponding to the engine RPM during the pass-by tests. The characteristic gearbox noise level is considered the result of the measurement of the noisiest gears at the noisiest side (left-right) of the gearbox. In 1993, a total number of 397 newly built units were tested and the measurement results are shown in Figure 7.21 in the form of the normal distribution probability plot, which is a graphical technique for assessing whether a data set is approximately normally distributed. The mean values and standard deviations are indicated in Figure 7.21. When discussion about the strategy for silencing gearboxes began, one of the suggestions recommended avoiding the extreme values of the gear deviations. Seven specimens with deviations within tolerances were selected for tests.

The empirical distribution function for 397 units is a brokenline (I and II sections), which indicates two subsets of the normally distributed data differing in the mean value and the standard deviation. It may be guessed that the origin of these subsets probably corresponds to the measurements of the gearboxes which differ significantly in quality, moreover, in two different ways. The slope of section I is less than the slope of the group of data with deviations within acceptable limits. The mean value of SPL determines the intersection of the horizontal line of 50% and the experimental distribution function. As is evident, the improvement of production quality is not sufficient for considerable reduction of noise. The SPL median of the checked gearboxes is less than the median of the others by approximately 0.8 dB. The only effect of selecting the gearbox is a reduction of the standard deviation of the SPL level values.

## 7.4    Gearbox Improvement Aimed at Noise Reduction

The research work on noise reduction employs different methods to identify the dynamic properties of the gearbox structure and to discover the dynamically weak parts. In reality, the effect of these improvements, which are limited by the given gearbox structure, on the overall SPL is of small significance. However, they are important because, after introducing these improvements, the SPL of newly built units varies in correlation with errors of gear geometry. The most efficient improvement can be reached when the noise problem is solved at its very source, which is the tooth contact of meshing gears [8].

## 7.4.1  Gearbox Housing Stiffness

The modal properties of the gearbox housing were identified using operational deflection shapes and experimental modal analysis. The effect of the gearbox load on the vibration response of each gear in the time domain using synchronous averaging with the rotational frequency was analysed. The 'mechanical power' can flow through the gearbox in many ways according to the gear used. The gearbox housing, shafts and gears are not an ideally rigid structure. The compliant gearbox structure is deformed under load. The toothmesh of the R-gear pair depends on the gear pair under load between the secondary and output shaft. The averaged vibration responses of the 21-tooth gear, belonging to the R-gear train, for the 3rd, 4th, and 5th gears under load are shown in Figure 7.22. The responses differ due to the gearbox deformation.

To prevent uncertain tooth contact, the gearbox housing was stiffened by using ribs, inside the housing the ribs were perpendicular to the shafts while two massive ribs at the housing surface were parallel to the shafts. The additional ribs ensure that the bearing close to the 21-tooth gear is stiffened enough and the vibration responses were not different [9].

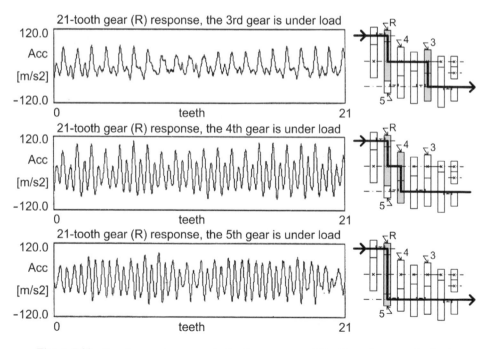

**Figure 7.22**   Synchronously averaged vibration response of the 21-tooth gear under load.

## 7.4.2   Geometric Design of Gears

The averaged toothmesh vibration signals are contact-ratio sensitive. It is well known that the contact ratio (simply, though not quite accurately stated, contact ratio is the average number of teeth in contact during a mesh cycle) is one of the most important parameters determining gear tooth excitation and, thus, gearbox noise level [10, 11, 12, 18, 19]. The effect of the tooth design on the average toothmesh which is calculated from the acceleration signal is shown in Figure 7.23. Measurement concerns helical gears under load whose design differs in profile contact ratio ($\varepsilon_\alpha$). The overlap contact ratio ($\varepsilon_\beta$ face contact ratio) is approximately equal to 1.0. The sum of the profile contact ratio and overlap ratio is designated as the total contact ratio ($\varepsilon_\gamma$).

The acceleration signal was measured on the input shaft bearing a direction perpendicular to the gear axis. The gear train marked by 'N' connects two parallel shafts, namely the input shaft and countershaft of the truck gearbox. The engagement of the gear train N or R splits the gear ratios 1 through 5 and doubles the number of the gear ratios to ten.

The average toothmesh of the gear train marked by N for the 3rd, 4th and 5th gear train under load, which is shown in Figure 7.23, demonstrates that the path of the mechanical power flow does not affect the response of this gear train. The average toothmesh of the mentioned gear train can be simply shown in Figure 7.24. The value of the profile contact ratio, which is less than 2.0, is called Low Contact Ratio (LCR) gearing while the gearing with this parameter equalled to 2.0 or more is designated as High Contact Ratio (HCR). The integer value for the profile contact ratio and the overlap ratio result in considerable reduction of gearbox vibration and noise. It can be estimated that introducing the HCR toothing results in reducing the noise level of gearboxes with LCR toothing by approximately 6 dB. As was already mentioned [2]

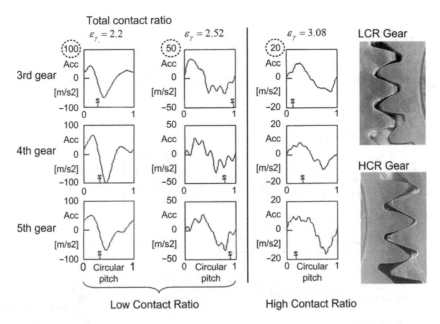

**Figure 7.23**   Effect of the total contact ratio on the average toothmesh acceleration signal of the N gear train for the 3rd, 4th and 5th gear train under load.

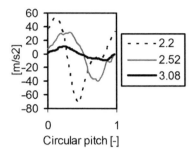

**Figure 7.24** Average toothmeshes vs. circular pitch for various total contact ratios.

the increase of the contact ratio from 4.0 to 5.0 decreases the transmission sound pressure level by approximately 10 dB.

The use of the averaged toothmesh signal is an effective method for verifying contact ratio, detecting a regular error in tooth profile geometry, and improving gearing by modification of tooth profile and lead while the envelope of acceleration signal during a complete revolution is important only for quality control.

The effect of tooth design on the gearbox noise level is shown in Figure 7.25, which represents the normal distribution probability plot containing three sets of data. The first one is a reference measurement of the gearbox noise with the LCR gears ($\varepsilon_\gamma = 2.52$) produced before 1994 originates from the tests of the gearboxes with geometric deviations within limits. The remaining sets of data correspond to the gearbox with the HCR gears ($\varepsilon_\gamma = 3.08$), production of which was started by TATRA in 1994, therefore, at the same time as for example in Germany

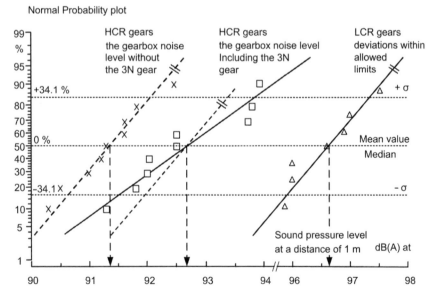

**Figure 7.25** Effect of LCR and HCR gears on experimental distribution functions of SPL.

(Ecosplit 3 type, ZF Friedrichshafen AG). The empirical distribution functions of these two measurements are computed including the measurement when the 3N gear is used and without this measurement. The intersection of the distribution function with the horizontal line of 50% results in the mean value (median for normally distributed data) of the gearbox SPL, which is indicated on the horizontal scale while the intersections with lines 50 ± 34.1% gives the doubled standard deviation. As one can note, the reduction of the noise level was by 4 dB and then after solution of the problem with the gear ratio of 3N even by 5 dB. The introduction of the tooth surface modification decreased the mean value of SPL below 90 dB as is shown in Figure 7.30.

## 7.5   Effect of Gear Quality on the Gearbox SPL

The gear tooth quality is given by the permissible maximum values of individual variations. Individual variations are those variations from their nominal values, which are exhibited by the various parameters of the gear teeth, such as pitch, profile shape, base diameters, pressure angle, tooth traces and helix angle. It is known that a gear could be absolutely perfect when referring to these variations, and yet be very noisy in the conditions of meshing. We can find in the tolerance data a few microns and have many times more deformations due to the loading. In reverse some low noisy gears are imperfect.

An example of the effect of the gear tooth quality on the emitted noise in SPL is shown in Figure 7.26. The data was taken from measurements on a truck gearbox with gears finished by grinding. It may be noticed that the SPL minimum (89 dB) corresponds to gears produced by a shape grinding machine (MAAG) while the others correspond to the grinding machine employing the continuous shift grinding method (Reishauer). However, it can be estimated that an improvement by class 1 of DIN 3961 quality results in the SPL reduction by approximately 1.5 dB. Gear Quality Class (GQC) is a quantity that is influenced mainly by the variations in the tooth profile (tooth trace form or tooth alignment angle) and weakly in the radial run-out. This quantity is computed as a weighted sum of the tooth trace form and tooth alignment angle (the value of the weighting coefficient for both deviations is equal to 0.4) and the radial run-out [13]. The gearbox noise level correlates with the gear geometry deviations if the gearbox structure is stiff enough.

The calculation of GQC as a real number was designed by Professor Moravec [11, 12, 13, 14], who studies the effect of individual deviations on gearbox noise. Individual deviations

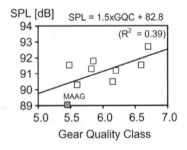

**Figure 7.26**   Effect of mean gear quality class on noise level in dB at 1 m.

are those deviations with nominal values which are exhibited by the various parameters of the gear teeth, such as circular pitch, profile shape, base diameters, pressure angle, tooth traces and helix angle. The formula for calculation of GQC contains a select set of the deviation types. The topic of this book is not a discussion on the significance of the deviation type, but the signal processing methods.

## 7.6    Effect of Operation Conditions on the Gearbox Vibrations

Increasing the torque at the gearbox input shaft causes misalignment due to the deformation of the gearbox case and the sytem of shafts including bearing. These deformations may result in shifts in the load distribution along and across the tooth surface of a gear pair which causes variation of the tooth contact stiffness as a main source of the parametric excitation of vibrations. Measurement of vibrations and the use of the special filtration technique can help in analysis of the dynamic force acting between teeth in mesh.

An example of the effect of gearbox load on the RMS value of the acceleration signal on the gearbox bearing is shown in Figure 7.27. The specimens of the gears with and without modified shape for testing were prepared by a supplier of grinders. The gearbox design input torque is equal to 1200 Nm. To prevent the increase of vibration and, consequently, the noise level due to the teeth and gearbox structure deformation, the modification of the tooth profile and lead by crowning across the face and tapering the lead is introduced [15]. The comparison of the modified toothing with the toothing, which is without modification, shows that the gear pair with the modified surface meshes more smoothly at nominal loading than the gears without modification. This design improvement can result in reducing SPL by 3 dB during the tests on the test stand.

The calculation of the average toothmesh can also be used as a tool to check the effect of the surface modification of the teeth on the overall level of vibration or noise. Total vibration values in Figure 7.28 correspond to average toothmesh in Figure 7.27.

The choice of appropriate surface modifications of the teeth can be considered art, which can handle only an experienced designer. Decisive evidence that the solution is optimal can only be obtained experimentally. Mathematical modelling is problematic because such parameters as the stiffness of the bearing is uncertain [15]. Suggesting a proper modification of the gear train at the beginning of the flow of mechanical power is easy for a small range of torque and

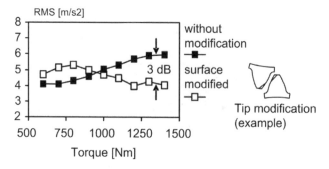

**Figure 7.27**    Effect of load and surface modification on RMS of acceleration.

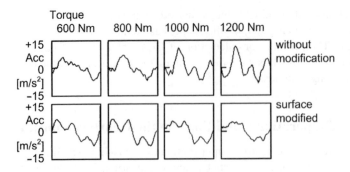

**Figure 7.28**   Effect of load and surface modification on average toothmesh.

speed. The difficulty of this task is for the gear train at the end of the power flow through the gearbox. This gear train operates in various conditions ranging from high speed rotation at small torque to low speed rotation at high torque. A useful tool for assessment of suitable modification may be SPL of the selected gear, which has been measured for different gear ratios, but with a constant torque at the input of the gearbox in Figure 7.29.

The positive effect of introducing the HCR gears into the TATRA gearboxes and the tooth surface modification on the noise level of 6 gear ratios (3R through 5N) which are important for the pass-by noise measurement is shown in Figure 7.30. The truck noise sources contribution to overall pass-by noise level for the nosiest gear (3N) is shown in Figure 7.16. It should be noted that the gear design of the tested truck was improved by introducing the HCR gears. The only implementation of HCR gears does not reduce noise transmission itself. It is necessary to improve the quality class at least by 1.5 to 2. The gearbox noise was not only reduced but the gearbox life time was doubled due to decreasing the dynamic forces acting between teeth. The aforementioned gearbox noise reduction caused some problems for the truck manufacturer when decreasing the SPL in the cab in such a way that drivers were able to distinguish the tonal noise emitted by the other unit, for example, by axles, mainly when they used engine

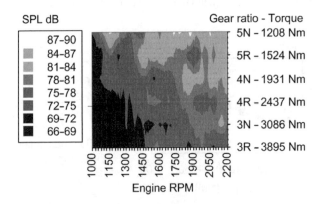

**Figure 7.29**   SPL of a gear train of the drop gearbox vs. torque at the main gearbox input.

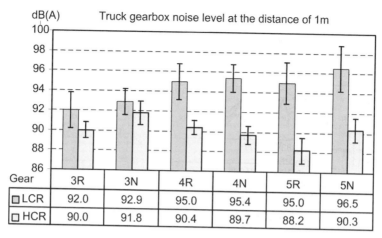

| Gear | 3R | 3N | 4R | 4N | 5R | 5N |
|------|------|------|------|------|------|------|
| ☐ LCR | 92.0 | 92.9 | 95.0 | 95.4 | 95.0 | 96.5 |
| ☐ HCR | 90.0 | 91.8 | 90.4 | 89.7 | 88.2 | 90.3 |

**Figure 7.30** The gearbox overall noise level of the HCR gears in comparison with the LCR gears for six engaged gears important for the pass-by noise level measurement.

braking. Cab environment is important from the point of view of the subjective perception of noise. The axle noise reduction is beyond the scope of this book.

It is important to note here that the gear of the HCR type has a positive influence on the lifetime of the transmissions. Smooth mesh reduces the amplitude of dynamic forces and the time interval up to the occurrence of fatigue fracture. The lifetime was approximately doubled.

## 7.7   Quality Control in Manufacturing

The production quality of a well designed gearbox, which reached excellent results in testing the prototype, should be carefully controlled. The gearboxes are tested during run-up/coast-down. The input shaft rotational speed was slowly increased from 1000 to 2200 RPM. The first five harmonics of the toothmeshing frequency is usually sufficient to set up the frequency range for measurements [16]. To demonstrate gear noise analysis, the dependence of the overall SPL in dB (Total) and the levels of five toothmeshing harmonics of all the gear trains under load on the input-shaft rotational speed for the 3N gears are shown in Figure 7.31. Except for the 5R gear, only three pairs of the engaged gears, which are marked by N, 3 and SG (Drop Gearbox), are under load. The panels of the diagram in Figure 7.26 titled Gear N, 3 and DG, corresponds to the already mentioned gear pairs. The curve in these panels marked by 'Sum', 'SumN', 'Sum3' and 'SumDG' are a sum of the power contributions of five harmonic components resulting in the noise level excited only by the appropriate pair of the gears. As the pass-by vehicle noise test is based on the maximum of the overall SPL, the maximum of the gearbox overall SPL (MaxTot) and a maximum of the five tonal components SPL (MaxSum) can be chosen as a gear quality criterion. Optionally, the maximum is evaluated for the input shaft rotational speed range either from 1000 to 2200 RPM or for an interval corresponding to the engine rotational speed during the pass-by tests. Due to the low rotational speed of the secondary gearbox gear train, its contribution to the overall (Total) SPL is negligible. The right lower panel in the diagram in Figure 7.31 compares the contribution of all the gear train under

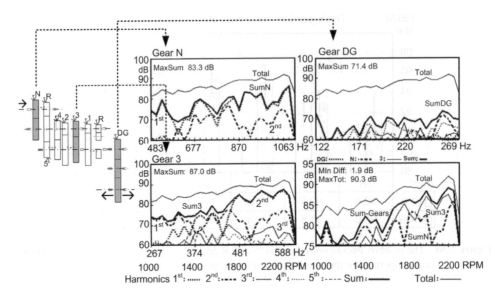

**Figure 7.31** Overall (Total) SPL and level of five toothmeshing harmonics of all the gear trains under load for the 3N gear vs. input-shaft rotational speed.

load to the overall SPL of the gearbox. The minimum of the difference between the overall SPL and the contributions of the N, 3 and SG gears for the already mentioned RPM range is designated by MinDiff. As was noted in the introductory section, the main sources of the gearbox noise are gears under load [9]. Other sources of noise, such as bearings, increase the SPL by 1.9 dB (MinDiff).

## References

[1] Munro, R.G. and Yildrim, N. (1994) Some measurements of static and dynamic transmission errors of spur gears. Proceedings of the International Gearing Conference 1994, University of Newcastle upon Tyne.

[2] Chung, Ch-H., Steyer, G., Abe, T. et al. (1999) Gear Noise Reduction Through Transmission Error Control and Gear Blank Dynamic Tuning, SAE Paper 1999-01-1766.

[3] Tůma, J. (2006) Simple gear set transmission error measurements. Proceedings of Thirteenth International Congress on Sound and Vibration (ICSV13), July 2–6, 2006, 8 p.

[4] Henriksson, M. and Pärssinen, M. (2003) Comparison of Gear Noise and Dynamic Transmission Error Measurements. Proceedings of Tenth International Congress on Sound and Vibration (ICSV10), Stockholm, pp. 4005–4012. (Paper 443)

[5] Smith, D.J. (1999) Gear Noise and Vibration, 1st edn, Marcel Dekker Inc., New York.

[6] Tůma, J. (2006) Dynamic transmission error measurement. Engineering Mechanics, 13(2), 101–106.

[7] Tůma, J. (2004) Dedopplerisation in Vehicle External Noise Measurements. Proceedings of Eleventh International Congress on Sound and Vibration (ICSV11), St. Petersburg, 5–8 July 2004, pp. 151–158.

[8] Tůma, J. (2009) Gearbox noise and vibration prediction and control. International Journal of Acoustics and Vibration, 14(2), 99–108.

[9] Tůma, J., Kuběna, R. and Nykl, V. (1994) Assessment of gear quality considering the time domain analysis of noise and vibration signals, in Proceedings of the 1994 International Gearing Conference, 1st edn, Technical University, Newcastle (UK), pp. 463–468.

[10] Houser, D. (2007) Gear noise and vibration prediction and control methods, Chapter 69, in *Handbook of Noise and Vibration Control* (ed. M. Crocker), Wiley, New York, 847–856.

[11] Hortel, M. and Škuderová, A. (2008) To the influence of nonlinear damping on the bifurcation phenomena in gear mesh of one branch of power flow of the pseudoplanetary gear system. Proceedings of Engineering Mechanics 2008, Czech Republic, Svratka.

[12] Welbourn, D.B. (1979) Fundamental knowledge of gear noise – A survey. Proceedings Noise & Vibration of Engineering and Transmission, I MECH E, Cranfield, pp. 9–14.

[13] Moravec, V. (1994) New toothing type in TATRA gearboxes with low noise and increased life time (in Czech). Proceedings of the International Conference ICESA 94, Praque, Czech Republic.

[14] Moravec, V. and Tůma, J. (2008) Metoda hodnocení jakosti ozubených soukolí měřením chyby převodu time (in Czech), Chapter 5, in *Identifikacja Stanu Dynamicznego i Trwalosci Przekladni Zebatych z Kolami o Uzebieniu Wysokim* (eds A. Skoc and M. Němček), Wydawnictwo politechniki Slaskiej, Gliwice, pp. 133–145.

[15] Dejl, Z. and Moravec, V. (2004) Modification of Spur Involute Gearing. Proceedings of The Eleventh World Congress on Mechanism and Machine Science, Tianjin, China, pp. 782–786.

[16] Tůma, J. (2007) Transmission and gearbox noise and vibration prediction and control, Chapter 88, in *Handbook of Noise and Vibration Control* (ed. M. Crocker), Wiley, New York, pp. 1080–1089.

[17] Tůma, J. and Moravec, V. (2000) Methods employed to assess the level of noise emitted by transmission units in relationship with the gear accuracy, in *Proceedings of Colloquium Dynamics of Machines 2000, National Colloquium with International Participation, February 8–9, 2000*, 1st edn, Institute of Thermomechanics AS CR, Prague, pp. 231–238.

[18] Drago, R.J. (1982) Gear system design for minimum noise, AGMA Handbook. (http://www.gear-doc.com/Publist.html)

[19] Drago, R.J. and Lenski, J.W. (1994) Overview of a five year research development and test programme. Proceedings of International Gearing Conference 1994, University of Newcastle upon Tyne.

# Index

*Vehicle Gearbox Noise and Vibration: Measurement, Signal Analysis, Signal Processing and Noise Reduction Measures,*
First Edition. Jiří Tůma.
© 2014 John Wiley & Sons, Ltd. Published 2014 by John Wiley & Sons, Ltd.